WORLD BANK INDEPENDENT EVALUATION GROUP

IEG

World Bank Assistance to Agriculture in Sub-Saharan Africa

An IEG Review

2007
The World Bank
Washington, D.C.

http://www.worldbank.org/ieg

1818 H Street NW
Washington DC 20433
Telephone: 202-473-1000
Internet: www.worldbank.org
E-mail: feedback@worldbank.org

1 2 3 4 5 10 09 08 07

Cover photo: A woman prepares the ground for planting, Mozambique. Photo by Eric Miller, courtesy of the World Bank.

ISBN-13: 978-0-8213-7350-7
e-ISBN-13: 978-0-8213-7351-4
DOI: 10.1596/978-0-8213-7350-7

World Bank InfoShop
E-mail: pic@worldbank.org
Telephone: 202-458-5454
Facsimile: 202-522-1500

Independent Evaluation Group
Knowledge Programs and Evaluation Capacity Development (IEGKE)
E-mail: eline@worldbank.org
Telephone: 202-458-4497
Facsimile: 202-522-3125

Printed on Recycled Paper

Contents

Figures

Harvesting rice in Senegal. Photo by Ray Witlin, courtesy of World Bank Photo Library.

Abbreviations

AAA	Analytical and advisory activities
AfDB	African Development Bank
ARD	Agriculture and Rural Development (Department, World Bank)
CAADP	Comprehensive Africa Agriculture Development Programme
CAE	Country Assistance Evaluation
CAS	Country Assistance Strategy
CDD	Community-driven development
CGAP	Consultative Group to Assist the Poorest
CGIAR	Consultative Group on International Agricultural Research
CIAT	International Center for Tropical Agriculture
CPIA	Country Policy and Institutional Assessment
DAC	Development Assistance Committee
DPL	Development policy lending
FAO	Food and Agriculture Organization of the United Nations
FARA	Forum for African Agricultural Research
FDI	Foreign direct investment
GCDS	Global Cassava Development Strategy
GDP	Gross domestic product
ICR	Implementation Completion Report
IDA	International Development Association
IEG	Independent Evaluation Group
IFAD	International Fund for Agricultural Development
IFDC	International Fertilizer Development Center
IFPRI	International Food Policy Research Institute
IITA	International Institute of Tropical Agriculture
ISNAR	International Service for National Agricultural Research
MDG	Millennium Development Goal
M&E	Monitoring and evaluation
NAADS	National Agricultural Advisory Services
NARS	National Agriculture Research Systems
NCPB	National Cereals and Produce Board
NEPAD	New Partnership for Africa's Development
NGO	Nongovernmental organization
O&M	Operations and maintenance
OECD	Organisation for Economic Co-operation and Development
OED	Operations Evaluation Department (now IEG)
ONCPB	Office National de Commercialisation des Produits de Base
OPCS	Operations Policy and Country Services (World Bank)
PAD	Project Appraisal Document
PER	Public Expenditure Review
PRSC	Poverty Reduction Support Credit
PRSP	Poverty Reduction Strategy Paper

QAG	Quality Assurance Group
R&D	Research and development
SOTOCO	Société Togolaise du Coton
T&V	Training and visit
UNDP	United Nations Development Programme
USAID	United States Agency for International Development
WDR	*World Development Report*

Acknowledgments

This report was prepared by a team consisting of Nalini Kumar (task manager), April Connelly, and Ridley Nelson under the guidance of Alain Barbu (manager, Independent Evaluation Group Sector, Thematic, and Global Evaluation [IEGSG]). Kavita Mathur and Tara Lonnberg provided research support. In addition to the core team, valuable contributions were received from Shawki Barghouti and Jumana Farah. Other IEG colleagues whose work provided input included Petros Aklilu, Chris Gerrard, Kieth Oblitas, and Keith Pitman. William Hurlbut edited the initial report and Caroline McEuen edited the manuscript for publication, Marie Charles provided administrative support.

The report benefited tremendously from the guidance provided by Ajay Chhibber, Sharoukh Fardoust, Patrick Grasso, and Vinod Thomas. Valuable comments were also received from Denis Carpio, Ken Chomitz, Victoria Elliott, Ali Khadr, James Sackey, and Klaus Tilmes of IEG. Gershon Feder peer reviewed the report.

Colleagues inside the Bank provided comments and input at various stages. They include: Jonathan Agwe, Ilhem Baghdadli, Mary A. Barton-Dock, Karen McConnell Brooks, Frank Fulgence K. Byamugisha, Derek R. Byerlee, Mark E. Cackler, Luc Christiaensen, Sanjiva Cooke, Christine E. Cornelius, IJsbrand Harko de Jong, Christopher Delgado, Ariel Dinar, Olivier Durand, Bleoue Nicaise Ehoue, Madhur Gautam, Steven Jaffee, Willem Janssen, Renate Kloeppinger-Todd, Patrick Labaste, Stephen Mink, Renato Nardello, David Nielson, Nwanze Okidegbe, Andrea Pape-Christiansen, Jeeva A. Perumalpillai-Essex, Christophe Ravry, Samjhana Thapa, Robert Townsend, John M. Underwood, and Martien Van Nieuwkoop. Several retired Bank staff provided valuable input based on their extensive experience, including Jock Anderson, Stephen Carr, Cornelis de Haan, Laurie Effron, Sushma Ganguly, Jacob Kampen, Satish Kumar, Uma Lele, and Jeffrey Lewis.

Director-General, Evaluation: *Vinod Thomas*
Director, Independent Evaluation Group, World Bank: *Ajay Chhibber*
Manager, IEGSG: *Alain Barbu*
Task Manager: *Nalini Kumar*

Hauling bananas by bicycle, Tanzania. Photo by Scott Wallace, courtesy of World Bank Photo Library.

Foreword

Sub-Saharan Africa is a highly complex and diverse Region that is a critical priority for the development community, as it has some of the world's poorest countries and remains behind on most of the Millennium Development Goals (MDGs). A major drag on Africa's development is the underperformance of the critical agriculture sector, which accounts for a large share of GDP and employment in the Region. This study assesses the development effectiveness of World Bank assistance in addressing constraints to agricultural development in Africa over the period of fiscal years 1991–2006.

The central finding of the study is that the agriculture sector has been neglected both by governments and the donor community, including the World Bank. The Bank's strategy for agriculture has been increasingly subsumed within a broader rural focus, in which its importance has suffered. Both arising from and contributing to this, technical skills to support agricultural development adequately have declined over time. The Bank's limited—and, until recently, declining—support for addressing the constraints on agriculture has not been strategically used to meet the diverse needs of a sector that requires coordinated intervention across a range of activities. The lending support from the Bank has been "sprinkled" across various agricultural activities such as research, extension, credit, seeds, and policy reforms in rural space, but with little recognition of the potential synergy among them to effectively contribute to agricultural development. As a result, though there have been areas of comparatively greater success—research, for example—results have been limited because of weak linkage with extension and limited availability of such complementary and critical inputs as fertilizers and water. Poor governance and conflict in several countries further complicate matters.

In order to effectively support the implementation of the Africa Action Plan and its appropriate focus on agricultural development as a key priority, the study has three recommendations for the Bank. First, it should focus attention on achieving improvements in agricultural productivity. Second, it should increase the quantity and quality of analytical work and ensure that policy advice and lending are grounded in its findings, and rebuild its technical skills. Third, it should establish clear benchmarks for measuring progress.

Vinod Thomas
Director-General, Evaluation

Avant propos

L'Afrique subsaharienne est une région très complexe et d'une grande diversité qui, parce qu'elle compte quelques-uns des pays les plus pauvres au monde et accuse un retard dans la réalisation de la plupart des objectifs de développement pour le Millénaire (ODM), est une priorité essentielle pour la communauté du développement. Les mauvais résultats du secteur névralgique de l'agriculture, qui est à l'origine d'une large part du PIB et de l'emploi dans la région, constituent un frein majeur au développement du continent. La présente étude évalue l'efficacité de l'aide apportée par la Banque mondiale pour remédier aux obstacles au développement agricole en Afrique pendant la période comprise entre les exercices 91 et 06.

L'étude aboutit à la conclusion fondamentale que le secteur agricole a été négligé tant par les autorités nationales que par la communauté des bailleurs de fonds, Banque mondiale comprise. La stratégie de la Banque pour l'agriculture a progressivement été intégrée à un projet rural de plus grande envergure où elle est passée au second plan. Les compétences techniques nécessaires à l'appui du développement agricole se sont en conséquence amenuisées au fil du temps, phénomène qui a à son tour entretenu cette situation. L'aide modérée _ et, jusque récemment, en repli _ apportée par la Banque mondiale pour remédier aux problèmes de l'agriculture n'a pas été utilisée de manière stratégique pour satisfaire aux besoins variés d'un secteur qui appelle des interventions coordonnées dans divers domaines. Les prêts consentis par la Banque ont été « dispersés » entre différentes activités agricoles, telles que la recherche, la vulgarisation, le crédit, les semences et les réformes de l'espace rural, mais sans guère tirer parti de leurs synergies éventuelles pour favoriser valablement le développement agricole. En conséquence, malgré la réussite comparative affichée dans certains domaines, comme la recherche, les résultats ont été limités en raison des liens ténus avec la vulgarisation et de la disponibilité insuffisante d'intrants complémentaires et critiques tels que les engrais et l'eau. La mauvaise gouvernance et les conflits qui sévissent dans plusieurs pays ne font que compliquer la situation.

Pour appuyer efficacement l'exécution du Plan d'action pour l'Afrique et de l'objectif prioritaire qu'il a judicieusement choisi, le développement agricole, l'étude formule trois recommandations à l'intention de la Banque mondiale. Elle doit d'abord axer ses efforts sur l'amélioration de la productivité agricole. Il lui faut ensuite augmenter la quantité et la qualité des études analytiques, veiller à ce que ses opérations de conseil et de prêt soient fondées sur ses observations, et restaurer ses compétences techniques. Enfin, elle doit établir des indicateurs précis pour mesurer les progrès.

Vinod Thomas
Directeur général, Évaluation

Prefácio

A África subsariana é uma Região muito complexa e diversa que constitui uma prioridade fundamental para a comunidade que se dedica ao desenvolvimento porque contém alguns dos países mais pobres do mundo e continua atrasada no que respeita a maioria dos Objectivos de Desenvolvimento para o Milénio (ODM). Um dos factores que atrasam o desenvolvimento da África é o fraco desempenho do importante sector da agricultura, o qual representa uma grande parte do PIB e do emprego na Região. O presente estudo avalia a eficácia para o desenvolvimento da assistência prestada pelo Banco Mundial ao abordar os constrangimentos ao desenvolvimento da agricultura na África durante o período compreendido entre os anos fiscais de 1991 e 2006.

A conclusão central do estudo é que o sector da agricultura tem sido negligenciado tanto pelos governos como pela comunidade de doadores, incluindo o Banco Mundial. A estratégia do Banco para a agricultura tem estado cada vez mais subordinada ao âmbito de uma focalização rural mais ampla, tendo perdido alguma da sua importância. Como resultado disso e contribuindo para isso, as competências técnicas para apoiar adequadamente o desenvolvimento da agricultura diminuíram com o tempo. O apoio limitado, e até gora decrescente, do Banco para abordar os constrangimentos à agricultura não tem sido utilizado estrategicamente para suprir as necessidades diversas de um sector que requer uma intervenção coordenada numa grande variedade de actividades. O apoio dos empréstimos do Banco tem sido "espalhado" por várias actividades agrícolas, tais como, investigação, extensão, crédito, sementes e reformas de política no espaço rural, mas a sinergia potencial entre eles para contribuir efectivamente para o desenvolvimento da agricultura pouco foi reconhecida. Consequentemente, embora em certas áreas tenha havido comparativamente mais êxitos, como por exemplo na investigação, os resultados foram limitados devido à debilidade das ligações com a extensão e à disponibilidade limitada de insumos complementares e fundamentais, como os fertilizantes e a água. A governação débil e os conflitos em vários países complicaram ainda mais a questão.

Para apoiar efectivamente a execução do Plano de Acção para a África e a sua focalização apropriada no desenvolvimento da agricultura como sendo uma prioridade principal, o estudo formula três recomendações ao Banco. Em primeiro lugar, o Banco deveria concentrar a sua atenção em obter melhoramentos na produtividade agrícola Em segundo lugar, o Banco deveria aumentar a quantidade e a qualidade dos seus trabalhos analíticos e assegurar que a assessoria de política e os empréstimos concedidos se baseiam nas suas conclusões, e reconstruir as suas competências técnicas. Em terceiro lugar, ele deveria estabelecer referências claras para medir os progressos alcançados.

Vinod Thomas
Director-Geral, Avaliação

Demonstration farm in Nigeria. Photo by Yosef Hadar, courtesy of World Bank Photo Library.

Preface

This IEG review of World Bank assistance to agriculture in Africa has a twofold purpose. First, it is a pilot for the proposed IEG study on Bank-wide assistance in agriculture scheduled for fiscal 2009. Second, the review provides timely insight into specific issues relevant to the Bank's renewed focus on agriculture in Africa, especially as expressed in the Africa Action Plan. In addition, the African Union has launched a vision and strategic framework for Africa's renewal—the New Partnership for Africa's Development (NEPAD). The Comprehensive Africa Agriculture Development Programme is at the heart of efforts by African governments under the NEPAD initiative to accelerate growth and eliminate poverty and hunger. Lessons of experience from the Bank will contribute to discussion surrounding these initiatives and will likely inform future international aid agendas and policy directions. The findings of the review also informed the Board of Directors' discussion of the *World Development Report 2008: Agriculture and Development.*

During the past two decades, the number of poor in Africa has doubled, from 150 million to 300 million. Africa remains behind on most of the Millennium Development Goals (MDGs) and is unlikely to reach them by 2015. About 70 percent of the target population for the MDGs is in the rural areas of Africa, and for most of those rural poor, agriculture is critical to successful attainment of the MDGs. Thus, an assessment of the Bank's contribution to agricultural development in the Region is critical to understanding the history of development of the sector and for drawing lessons for the future.

Préface

L'examen par l'IEG de l'assistance apportée par la Banque mondiale à l'agriculture en Afrique a une double finalité. Il s'agit en premier lieu d'une étude pilote préalable à l'analyse de l'aide de la Banque mondiale à l'agriculture en général, que l'IEG prévoit d'effectuer durant l'exercice 09. Il apporte ensuite des informations actualisées sur des questions particulières qui présentent un intérêt pour la Banque dans le cadre du recentrage de ses activités sur l'agriculture en Afrique, tel qu'exprimé dans le Plan d'action pour l'Afrique notamment. Par ailleurs, l'Union africaine a formulé un projet et un cadre stratégique pour le renouveau de l'Afrique – le Nouveau partenariat pour le développement de l'Afrique (NEPAD). Le Programme détaillé pour le développement de l'agriculture africaine est au centre des efforts déployés par les gouvernements africains dans le cadre du NEPAD en vue d'accélérer la croissance et d'éliminer la pauvreté et la faim. Les enseignements dégagés de l'expérience de la Banque mondiale alimenteront le débat entourant ces projets et seront de nature à inspirer les programmes d'aide internationaux et les orientations stratégiques futurs. Les conclusions de l'examen ont également guidé la discussion du Conseil des Administrateurs portant sur le Rapport sur le développement dans le monde 2008 : L'agriculture au service du développement.

Au cours des deux dernières décennies, le nombre de pauvres a doublé en Afrique, passant de 150 millions à 300 millions. L'Afrique accuse un retard dans la réalisation de la plupart des objectifs de développement pour le Millénaire (ODM) et a peu de chance de les atteindre d'ici à 2015. Quelque 70 % de la population ciblée par les ODM vit en milieu rural et, pour la plupart des habitants pauvres des zones rurales, l'agriculture est un moyen capital d'atteindre des ODM. Il est donc indispensable de procéder à une évaluation du concours de la Banque mondiale au développement agricole de la région, pour comprendre l'histoire du développement de ce secteur et en dégager des enseignements pour l'avenir.

Nota Preliminar

A presente análise do IEG da assistência do Banco Mundial à agricultura na África tem um duplo propósito. Primeiro, é uma análise piloto para o estudo proposto pelo IEG sobre a assistência prestada pela totalidade do Banco à agricultura, programado para o ano fiscal de 2009. Segundo, a análise proporciona uma visão em tempo oportuno de questões específicas que são pertinentes à focalização renovada do Banco da agricultura na África, especialmente como foi expressa no Plano de Acção para a África. Adicionalmente, a União Africana adoptou uma visão e estrutura estratégica para a renovação da África — a Nova Parceria para o Desenvolvimento da África (NEPAD). O Programa Abrangente para o Desenvolvimento da Agricultura na África está no âmago dos esforços realizados pelos governos africanos, no âmbito da iniciativa do NEPAD, para acelerar o crescimento e eliminar a pobreza e a fome. Os ensinamentos obtidos com a experiência do Banco vão contribuir para o debate em torno destas iniciativas e provavelmente informarão as agendas das ajudas internacionais e a direcção das medidas de política. As conclusões da análise também informaram as deliberações do Conselho de Administração relativas ao *Relatório sobre o Desenvolvimento Mundial* de 2008: A Agricultura e o Desenvolvimento.

Durante as últimas duas décadas, o número de pobres na África duplicou, passando de 150 milhões para 300 milhões. A África continua a ficar para trás no que respeita a maioria dos Objectivos de Desenvolvimento para o Milénio (ODM) e é improvável que os alcance até ao ano de 2015. Cerca de 70 por cento da população visada pelos ODM encontram-se nas zonas rurais da África, e para a maioria destes pobres das zonas rurais a agricultura é crucial para conseguir alcançar os ODM. Assim, uma avaliação da contribuição do Banco para o desenvolvimento da agricultura na Região é fundamental para compreender a história do desenvolvimento deste sector e para obter ensinamentos para o futuro.

Dinka herders care for their cattle near Mapourit, Sudan. Photograph by Douglas Engle and reproduced with his permission.

Executive Summary

Sub-Saharan Africa is a highly complex Region of 47 countries with 7 distinctly different colonial histories. It is also highly diverse, with more than 700 million people and at least 1,000 different ethnic groups. The Region is a critical development priority. It includes some of the world's poorest countries, and during the past two decades the number of poor in the Region has doubled to 300 million—more than 40 percent of the Region's population. Africa remains behind on most of the Millennium Development Goals (MDGs) and is unlikely to reach them by 2015.

A major drag on Africa's development is the underperformance of the agriculture sector. This is a critical sector in the Region, because it accounts for a large share of gross domestic product (GDP) and employment. The weak performance of the sector stems from a variety of constraints that are particular to agriculture in Africa and make its development a complex challenge. Poor governance and conflict in several of the countries further complicate matters. IEG has assessed the development effectiveness of World Bank assistance in addressing constraints to agricultural development in Africa over the period of fiscal years 1991–2006 in a pilot for a wider assessment of the Bank's assistance to agriculture worldwide.

The central finding of the study is that the agriculture sector has been neglected by both governments and the donor community, including the World Bank. The Bank's strategy for agriculture has been increasingly subsumed within a broader rural focus, which has diminished its importance. Both arising from and contributing to this, the technical skills needed to support agricultural development adequately have also declined over time.

The Bank's limited—and, until recently, declining—support for addressing the constraints on agriculture has not been used strategically to meet the diverse needs of a sector that requires coordinated intervention across a range of activities. The lending support from the Bank has been "sprinkled" across various agricultural activities such as research, extension, credit, seeds, and policy reforms in rural space, but with little recognition of the potential synergy among them to effectively contribute to agricultural development. As a result, though there have been areas of comparatively greater success—research, for example—results have been limited because of weak linkage with extension and limited availability of such complementary and critical inputs as fertilizers and water. Hence the Bank has had limited success in contributing to the development of African agriculture.

The Challenges of African Agriculture

Agricultural output has grown in Africa, but it is difficult to calculate a reliable growth rate for the Region over the study period because of wide variations across countries and over time. Some countries, such as Gabon, moved from poor performance in 1990–2000 to better performance in 2000–04; others, such as Malawi, moved in the opposite direction. The change has often been dramatic, which makes aggregate growth rates misleading. For example, agriculture in Angola grew at 13.7 percent a year during

2000–04, although growth had retreated by 1.4 percent yearly during 1990–2000. Only about a quarter of the countries in the Region, among them Benin, Burkina Faso, Ghana, Nigeria, and Tanzania, show consistent agricultural growth of over 3 percent in the 1990–2004 period.

Total agricultural output in Africa consists primarily of food crops. Agricultural export crops account for less than 10 percent of total production. While some export crops, including cotton, have contributed to poverty alleviation in countries such as Burkina Faso, food crops have performed poorly in most countries. Cereal yields in Africa, even in 2003–05, were less than half those in South Asia and one-third those in Latin America. Africa also lags behind other Regions in the percentage of cropland irrigated, fertilizer use, and labor and land productivity per worker. While the great strides in South Asia's agricultural production from 1961 to 2001 were mainly the result of increased yields, gains in food production in Africa were produced primarily through the expansion of cultivated land. Meanwhile, crop yields stagnated.

Beginning in 1973, Africa became a net food importer. Since that time, food production has not kept pace with the rapidly growing population, and food imports have grown rapidly. Meanwhile, Africa's exports, which are primarily agriculture-based, declined; for several commodities, including coffee, the Region's share of the world market evaporated. Agricultural subsidies in Organisation for Economic Co-operation and Development (OECD) countries have played a major role in keeping world prices low for several of these crops. This, among other factors, has impacted the adequacy of returns to farmers.

Agriculture in Africa is primarily a family activity, and the majority of farmers are smallholders who own between 0.5 and 2.0 hectares of land, as determined by socio-cultural factors. Women provide about half of the labor force and produce most of the food crops consumed by the family.

Agricultural land in Africa falls into several agro-ecological zones that run across countries. It is largely characterized by poor soils, highly variable rainfall, and frequent droughts. Transport infrastructure is poor, access to irrigation is limited, and under rain-fed conditions, chronic food insecurity is a reality for millions of small farmers. To survive in this harsh environment, most farmers rely on diversified coping strategies. To ensure at least some produce from their land, African farmers normally plant several varieties of crops (typically 10 or more) with different maturation periods, together with trees. Livestock is also an important source of security for farmers in Africa, particularly in lean years. The average smallholder's access to credit is also extremely limited. Hardy crops such as millet, sorghum, cassava, and other root crops are more important than cereals such as rice and wheat, which were the mainstay of the Asian Green Revolution.

In this environment, for farmers to have an incentive to practice intensive agriculture and take risks with new crop varieties, a number of factors need to come together at the same time, or at least appear in an optimal sequence, including improved seeds, water, credit, and access to markets; good extension advice; and adequate returns through undistorted prices for inputs and outputs. A strategy for development of agriculture in Africa must consider each of these factors in the context of Africa's unique characteristics and specific local conditions.

Past Approaches to African Agriculture

Until very recently, agricultural development in Africa was neglected by both governments and donors. During the 1960s, immediately following independence, governments in several African countries considered agriculture primarily a source of resources for industrialization. Then, in the 1970s, the World Bank led the shift toward a broader development model in Africa that was consistent with a more general shift in the understanding of development. This committed the institution to integrated rural development to directly attack Africa's rural poverty and underdevelopment. In the mid-1980s, when African countries faced severe fiscal crises, donors prioritized improvements in the efficiency of resource allocation and pressed agriculture marketing reforms. But structural reforms also fell short of producing the desired growth effects.

The Role of Aid

Bilateral and multilateral donor aid for development of African agriculture declined from $1,921 million in 1981 to $997 million in 2001 (in 2001 dollars). Lending from both sources has since rebounded with the increasing focus on African development. OECD data show that although bilateral donors as a group have played a comparatively larger role, the World Bank was the single largest donor to African agriculture between 1990 and 2005. The largest bilateral donors were the United States and Japan.

Foreign private sector flows into Africa are modest in comparison with bilateral and multilateral aid (Hazell and von Braun 2006). Private commercial investment in African agriculture has been largely limited to export crops and higher-potential zones. A number of international seed companies have invested in maize seed multiplication, and in September 2006 the Rockefeller and Bill and Melinda Gates Foundations together launched a new partnership to help Africa develop its agriculture.

Agriculture's Potential and the Bank's Strategy

For Africa to meet the MDGs, it will be necessary to realize the potential of the agriculture sector, to provide the support needed for it to contribute to growth and poverty reduction. Research by Dorosh and Haggblade (2003) and IFPRI (2006a) found that investments in agriculture generally favor Africa's poor more than similar investments in manufacturing.

The World Bank has not had a separate strategy for agriculture in Africa except as part of its wider rural development strategies, and over time the agriculture strategy was subsumed in a broader rural focus. More recently, however, the Africa Action Plan has recognized the agriculture sector as a potential driver of growth.

The Bank's Overall Assistance and Its Assessment

Over fiscal years 1991–2006, the Bank provided the countries of the Africa Region with $2.8 billion in investment lending (as distinct from adjustment lending) in agriculture, constituting 8 percent of total Bank investment lending to the Region. A large part of this lending has been in the form of agriculture components in rural projects. In addition, there have been 77 Development Policy Loans with agriculture components, and in 18 of these, agriculture was a significant dimension.

This limited investment lending has performed below par. IEG data show that the percentage of satisfactory outcome ratings for largely agricultural investment projects during 1991–2006 is lower than that for non-agriculture investments in the Region (60 against 65 percent satisfactory). It is also lower than the percentage for similar investment projects in other Bank Regions (73 percent satisfactory). Sustainability ratings are also below average. Although further analysis is needed, the study found that largely agricultural projects in countries with less favorable agricultural conditions have done better than similar projects in countries with more favorable conditions.

The Bank's activities in support of agricultural development in Africa have comprised lending, analytical work, and policy advice. Until very recently the analytical work—necessary for the diagnosis of issues and actions and to help shape the policy advice and lending—has been limited, scattered, of variable quality, and not easily available. In addition, IEG found that there are no specific procedures in place to ensure that the findings of analytical work are systematically reflected in lending and policy dialogue.

IEG found that the lending support provided by the Bank has not reflected the interconnected nature of agriculture activities. Rather, the lending has been "sprinkled" across an array of activities in rural space, including research, extension, marketing reform, drought relief, seed development, and transport, but with little recognition of the relationships among them and the need for all of these areas to be developed at the same time, or at least in an optimal sequence, to effectively contribute to agricultural development. While the Bank's broader rural focus from the mid-1980s was justified, an unintended result was that it led to less focused attention on the need for various activities that are critical for agricultural development in rural space to come together at the same time or to take place in some optimal sequence.

This review found that none of the top 10 borrowers, among them Côte d'Ivoire, Ethiopia, Tanzania, and Uganda, had received *consistent* and *simultaneous* support across all critical subsectors. That is not to suggest that the Bank should do this alone—it might well be done better in partnership—but the Bank could reasonably be expected to take the lead in fostering such a multifaceted approach, based on its comparative advantage as a multisector lending institution.

Thematic Performance

An assessment of the achievements and shortcomings in the Bank's support by main theme reveals a mixed record:

Agro-ecological diversity. Bank support has helped build the capacity of national research systems and develop zonal stations to give an agro-ecological focus to research. However, there is little indication that Bank projects other than research interventions have systematically adapted activities to diverse country agro-ecological conditions. The ability to respond to local conditions has been the primary appeal of projects that use community-based approaches, but there is little evidence that these approaches, as used in projects in Ghana and Tanzania, for example, are able to respond to agro-ecological diversity.

Fluctuating rainfall and droughts. Bank projects completed through fiscal 2006 have been responsive to drought emergencies and have helped governments set up drought management systems. But they have not been able to help countries such as Malawi, for example, develop a long-term strategic approach to address the basic factors that create food insecurity—that is, to help countries increase agricultural productivity sufficiently to arrest declining per capita food availability. In this connection, while the Bank has contributed to development of improved millet and cassava varieties through support to research, it has missed the opportunity to recognize the important role that cassava can play in promoting food security in most countries.

Poor soil fertility. The Bank has been party to several international and regional initiatives on this issue, including the *Terr Africa Regional Initiative,* launched in 2005. This multidimensional partnership is expected to promote a collective approach to sustainable land management in the Region. But Bank lending appears to have addressed soil fertility more as an environmental than as an agricultural productivity issue.

Access to water. Though the Bank has identified the need for investment in irrigation, it has done very limited lending for that purpose. The Bank interventions that support water management in rain-fed areas have achieved physical targets, but because of poor monitoring and evaluation (M&E), it is difficult to tell what has worked and what has not.

Improved seeds. The Bank has contributed to the Consultative Group on International Agricultural Research (CGIAR), which has made significant contributions in this area, and Bank projects have also provided opportunity for testing and scaling up technologies, as in Ethiopia and Togo. Nonetheless, seed-related activities have so far made only a modest contribution to increases in crop production. Bank projects have also not been able to address the issue of limited use of seeds by farmers because of inadequate access to complementary inputs.

Farmers' access to credit and rural finance. Overall support from the Bank in this critical area has been limited. Aside from institutional capacity weaknesses in client countries, one reason for this low level of support has been weak project performance in this area, brought about by, among other things, weak implementation of Bank guidelines, particularly regarding eligibility and performance of financial intermediaries.

There is need for the Bank to take greater care in designing and supervising these operations, and all options should continue to be explored for the most appropriate way to provide farmers with the means necessary to increase productivity and incomes.

Poor transport infrastructure. Bank-supported agriculture interventions have made only a limited contribution to improving transport infrastructure to promote market access for agricultural development.

Weak extension. The Bank has helped raise client awareness about the importance of extension to agricultural development. It currently supports a range of partnership approaches (public-private, demand-driven, nongovernmental organizations, and so on), as in Uganda. But the cost, effectiveness, and sustainability of these approaches need to be systematically evaluated.

Price and marketing reform. Though results have been variable across countries, the Bank's effort has contributed to improving the macroeconomic environment and fiscal discipline in several countries. However, these changes were not enough to stimulate private sector investments in several critical areas from which the public sector withdrew. Consequently, most countries in Africa face exorbitant fertilizer prices, inadequate seed production, poor transport, and limited credit access. While the reform process had limited positive impact on food production, it nevertheless boosted production of nontraditional export crops such as mangoes from Mali and flowers from Kenya. Beyond individual countries, the Bank lobbied for a genuinely pro-development Doha Round and for elimination of OECD agricultural subsidies in international forums, but with limited success to date.

Insecurity of tenure. Analytical work has contributed to a better understanding of property rights regimes. But the Bank has found it difficult to provide effective support in this area because of its political, social, and cultural sensitivity.

The Millennium Development Project Hunger Task Force concluded in 2005 that the world could meet the MDG of halving hunger by 2015. Development of African agriculture is critical to achieving this goal, and the World Bank can make a major contribution because it is one of the largest sources of development finance for agriculture and can provide policy advice to governments.

Key Findings on Bank and Country Factors of Performance

Bank factors

- The institution's strategy for the development of the agriculture sector has been part of its rural strategy, and over time the importance of agriculture in the Bank's rural strategy has declined. Both arising from and contributing to this, technical skills to support agricultural development adequately have also declined over time. Data from the Human Resources Department of the World Bank show that there were 17 technical experts mapped to the Agriculture and Rural Development Department in Sub-Saharan Africa in 2006, compared with 40 in 1997.
- The Bank's diagnosis of a country's development status and priorities in the agriculture sector is carried out primarily through analytical work. Until very recently this work has been limited and not readily available. Nor have the findings from analytical work strategically informed Bank–client policy dialogue and lending program design.
- Bank policy advice appears to have had far-reaching implications for the direction of agricultural development in African countries, in particular its policy advice associated with the adjustment agenda. However, results have fallen short of expectations because of weak political support and insufficient appreciation of reality on the ground, among other things.
- The Bank's data systems and support for M&E have been insufficient to adequately inform the institution's effort to develop agriculture in Africa across a broad front. Current data systems do not allow the institution to track in

enough detail how much is being provided for development of specific activities such as seed development and credit. M&E at the project level has been of limited value in answering fundamental questions about outcome, impact, and efficiency, such as who benefited, which crops received support and how, what has been the comparative cost effectiveness, and to what can one attribute gains.

Country factors

- Although the governance environment in several African countries continues to be weak, political commitment for the development of agriculture in client countries appears stronger than in the past. African governments, many of which were allocating less than 1 percent of their budget to agriculture, agreed in July 2003 at the African Union Summit to allocate at least 10 percent of national budgetary resources for programs to support agricultural growth in the next five years.
- Considerable agricultural research capacity exists, although the sustainability of the activities supported remains uncertain. Overall, government capacity in several countries remains weak, and local agriculture ministries are still relatively ineffective partners in promoting development of the agriculture sector. Though further analysis is needed, the study finding that largely agricultural projects in countries with less favorable agricultural conditions have done better than similar projects in countries with more favorable conditions suggests that other factors—such as political economy and country capacity—are also a challenge for agricultural development in Africa.

Recommendations

To effectively support the implementation of the Africa Action Plan and its appropriate focus on agricultural development as a key priority, IEG recommends that the Bank:

1. Focus attention to achieve improvements in agricultural productivity:
   - Establish realistic goals for expansion of irrigation and recognize the need to increase productivity of rain-fed agriculture through improvements in land quality, as well as water and drought management.
   - Help design efficient mechanisms, including public-private partnerships, to provide farmers with critical inputs, including fertilizers, water, credit, and seeds.
   - Support the development of marketing and transport infrastructure.
2. Improve its work on agriculture:
   - Increase the quantity and quality of analytical work on agriculture and ensure that policy advice and lending are grounded in its findings.
   - Support public expenditure analyses to assess resource availability for agriculture and to help set Bank priorities.
   - Rebuild its technical skills, based on a comprehensive assessment of current gaps.
3. Establish benchmarks for measuring progress:
   - Improve data systems to better track activities supported by the Bank.
   - Strengthen M&E to report on project activities in various agro-ecological zones and for different crops and farmer categories, including women.
   - Develop a system to coordinate agricultural activities in a country with road access, market proximity, and soil conditions.

Synthèse

L'Afrique subsaharienne est une région très complexe composée de 47 pays ayant connu sept histoires coloniales distinctes. Elle est également très diversifiée puisqu'elle compte 700 millions d'habitants issus d'au moins un millier de groupes ethniques. Le développement de l'Afrique est une priorité stratégique. Le continent compte quelques-uns des pays les plus pauvres au monde. Au cours des vingt dernières années, le nombre de pauvres y a doublé pour atteindre 300 millions — soit plus de 40 % de la population totale. L'Afrique accuse un retard dans la réalisation de la plupart des objectifs de développement pour le Millénaire (ODM) et a peu de chances de les atteindre d'ici à 2015.

Les mauvais résultats du secteur de l'agriculture constituent l'un des principaux freins au développement de l'Afrique. Ce secteur revêt une importance de premier plan pour la région du fait qu'il représente une part notable du produit intérieur brut (PIB) et de l'emploi. Sa mauvaise performance tient à différents problèmes qui sont particuliers à l'agriculture africaine et font de son développement un enjeu complexe. La mauvaise gouvernance et les conflits qui sévissent dans plusieurs pays ne font que compliquer la situation. L'IEG a mesuré l'efficacité de l'aide apportée par la Banque mondiale en vue de résoudre les obstacles au développement agricole en Afrique entre les exercices91 et 06 dans le cadre d'une étude pilote préalable à une évaluation plus générale de l'assistance qu'elle fournit à l'agriculture à l'échelle mondiale.

La conclusion essentielle de l'étude est que le secteur agricole a été négligé tant par les gouvernements que par la communauté des bailleurs de fonds, Banque mondiale comprise. La stratégie de la Banque pour l'agriculture a progressivement été intégrée à un objectif rural de plus grande envergure où elle est passée au second plan. Les compétences techniques nécessaires à l'appui du développement agricole se sont en conséquence amenuisées au fil du temps, phénomène qui a lui-même entretenu cette situation.

L'aide modérée – et, jusque récemment, en repli – apportée par la Banque mondiale pour remédier aux problèmes de l'agriculture n'a pas été utilisée de manière stratégique pour satisfaire aux besoins divers d'un secteur qui appelle des interventions coordonnées dans divers domaines. Les prêts consentis par la Banque ont été « dispersés » entre différentes activités agricoles, telles que la recherche, la vulgarisation, le crédit, les semences et les réformes de l'espace rural, mais sans guère tirer parti de leurs synergies éventuelles pour favoriser valablement le développement agricole. En conséquence, malgré la réussite comparative affichée dans certains domaines, comme la recherche, les résultats ont été limités en raison des liens ténus avec la vulgarisation et de la disponibilité insuffisante d'intrants complémentaires et critiques tels que les engrais et l'eau. La Banque mondiale n'est donc pas parvenue véritablement au développement de l'agriculture africaine.

Les problèmes de l'agriculture africaine

La production agricole africaine a progressé, mais il est difficile de calculer un taux qui rende réellement compte de cette progression à l'échelle de la région pour la période couverte par l'étude, en raison des disparités importantes entre les pays et des variations dans le temps. Dans certains pays, comme le Gabon, les résultats médiocres enregistrés entre 1990 et 2000 se sont redressés entre 2000 et 2004. D'autres, comme le Malawi, ont connu une évolution inverse. La mutation a souvent été spectaculaire, ce qui fait que les taux de croissance agrégés sont trompeurs. En Angola, par exemple, l'agriculture a progressé de 13,7 % par an entre 2000 et 2004, alors qu'elle avait reculé de 1,4 % par an entre 1990 et 2000. Un quart seulement des pays de la région, dont le Bénin, le Burkina Faso, le Ghana, le Nigéria et la Tanzanie, affichent une croissance agricole régulière de plus de 3 % pour la période comprise entre 1990 et 2004.

Les cultures vivrières constituent l'essentiel de la production agricole totale de l'Afrique. Les cultures d'exportation en représentent moins de 10 %. Si certaines cultures d'exportation, dont le coton, ont concouru à l'allègement de la pauvreté dans des pays tels que le Burkina Faso, le rendement des cultures vivrières a été médiocre presque partout. Même pendant la période comprise entre 2003 et 2005, les rendements céréaliers de l'Afrique ont été inférieurs de plus de moitié à ceux de l'Asie du Sud et d'un tiers à ceux de l'Amérique latine. L'Afrique est également en retard par rapport à d'autres régions en termes de pourcentage de terres agricoles irriguées, d'utilisation d'engrais, et de productivité du travail et de la terre par travailleur. Alors que les progrès remarquables de la production agricole de l'Asie du Sud entre 1961 et 2001 ont essentiellement été dus à la hausse des rendements, l'augmentation de la production alimentaire en Afrique a principalement tenu à l'expansion des terres cultivées. Pendant ce temps, les rendements agricoles ont stagné.

En 1973, l'Afrique est devenue importatrice nette de produits alimentaires. Par la suite, la production alimentaire n'a pu s'adapter à la flambée démographique, et les importations ont rapidement augmenté. Dans le même temps ses exportations, essentiellement composées de produits agricoles, ont fléchi ; sa part du marché mondial pour plusieurs produits de base, dont le café, s'est volatilisée. Les subventions agricoles des pays membres de l'Organisation de coopération et de développement économiques (OCDE) ont puissamment contribué à maintenir le prix de plusieurs de ces produits à un bas niveau ce qui, entre autres facteurs, a pénalisé la rémunération des agriculteurs.

En Afrique, l'agriculture est principalement une activité familiale. La majorité des agriculteurs sont de petits exploitants qui possèdent entre 0,5 et 2 hectares de terre, selon les facteurs socioculturels. Les femmes constituent environ 50 % de la main d'œuvre et produisent l'essentiel des cultures vivrières consommées par la famille.

Les terres agricoles se répartissent entre plusieurs zones agroécologiques qui couvrent différents pays. La pauvreté des sols, la forte variabilité des précipitations et la fréquence élevée des sécheresses en sont les principales caractéristiques. L'infrastructure de transport est insuffisante, l'accès à l'irrigation limité et, en cas d'agriculture pluviale, l'insécurité alimentaire chronique est une réalité pour des millions de petits agriculteurs. Pour survivre dans cet environnement hostile, la plupart des agriculteurs ont recours à des stratégies d'adaptation diversifiées. Pour être certains d'obtenir un minimum de produits de leur terre, ils plantent habituellement plusieurs variétés végétales (une dizaine au moins en général) ayant des cycles de maturation différents, ainsi que des arbres. Le bétail est aussi pour eux une source importante de sécurité, surtout en période difficile. L'accès au crédit du petit exploitant type est aussi extrêmement restreint. Les cultures rustiques, comme le millet, le sorgho, le manioc et d'autres plantes à racines, sont plus importantes que des céréales telles que le riz et le blé, qui ont été à la base de la Révolution verte en Asie.

Dans ce contexte, pour inciter les agriculteurs à

pratiquer une agriculture intensive et à prendre le risque de semer de nouvelles variétés, plusieurs facteurs doivent être réunis, ou tout du moins s'enchaîner de manière optimale, à savoir : l'accès à des semences des semences améliorées, à l'eau, à des crédits et aux marchés ; de judicieux conseils de vulgarisation ; et des rendements suffisants, en l'absence de distorsion des prix des intrants et des extrants. Une stratégie de développement de l'agriculture en Afrique doit tenir compte de chacun de ces facteurs dans le cadre des particularités propres au continent et des situations locales spécifiques.

L'agriculture africaine : Approches antérieures

Jusque très récemment, le développement agricole de l'Afrique a été négligé tant par les autorités nationales que par les bailleurs de fonds. Dans les années 60, tout de suite après l'indépendance, les gouvernements de plusieurs pays africains voyaient avant tout dans l'agriculture un moyen de produire des ressources pour l'industrialisation. Plus tard, dans les années 70, la Banque mondiale a été la première à adopter pour le continent un modèle de développement plus large, qui s'inscrivait dans l'évolution plus générale de la théorie du développement. L'institution a ainsi adhéré au développement rural intégré pour s'attaquer de front à la pauvreté et au sous-développement ruraux en Afrique. Au milieu des années 80, alors que les pays africains étaient confrontés à de graves crises budgétaires, les bailleurs de fonds ont donné priorité à l'allocation plus efficiente des ressources et ont appelé à des réformes du système de commercialisation des produits agricoles. Les réformes structurelles n'ont cependant pas permis de produire les résultats souhaités en termes de croissance.

Le rôle de l'aide

Entre 1981 et 2001, l'aide bilatérale et multilatérale au développement de l'agriculture africaine a diminué, passant de 1 921 millions de dollars à 997 millions de dollars (en dollars de 2001). Les prêts émanant de ces deux sources ont augmenté depuis lors suite au recentrage des priorités sur le développement de l'Afrique. Les données de l'OCDE montrent que si la contribution globale des bailleurs de fonds bilatéraux a été comparativement supérieure, la Banque mondiale a été le principal donateur à l'agriculture africaine entre 1990 et 2005, les plus gros bailleurs de fonds bilatéraux étant les États-Unis et le Japon.

Les apports de capitaux étrangers privés sont modestes en comparaison aux apports bilatéraux et multilatéraux (Hazell et von Braun 2006). L'investissement commercial privé dans l'agriculture africaine s'est en grande partie cantonné aux cultures d'exportations et aux zones présentant un meilleur potentiel. Plusieurs entreprises semencières internationales ont investi dans la multiplication de semences de maïs et, en septembre 2006, la fondation Rockefeller et la Fondation Bill et Melinda Gates ont donné le coup d'envoi à un nouveau partenariat pour aider l'Afrique à développer son agriculture.

Le potentiel de l'agriculture et la stratégie de la Banque mondiale

Pour que l'Afrique atteigne les ODM, il faudra réaliser le potentiel du secteur agricole et prêter l'assistance nécessaire pour que celui-ci concoure à la croissance et à la lutte contre la pauvreté. Les études de Dorosh et Haggblade (2003) et de l'IFPRI (2006a) ont constaté que les investissements dans l'agriculture sont généralement plus bénéfiques aux pauvres que des investissements équivalents dans l'industrie manufacturière.

La Banque mondiale n'a pas adopté de stratégie particulière pour l'agriculture en Afrique, si ce n'est dans le cadre de ses stratégies globales de développement rural et, avec le temps, sa stratégie agricole a été intégrée à un projet rural de plus grande ampleur. Plus récemment, toutefois, le Plan d'action pour l'Afrique a pris le potentiel de croissance du secteur agricole en considération.

L'assistance globale de la Banque mondiale et son évaluation

Entre les exercices 91 et 06, la Banque mondiale a consenti aux pays de la région Afrique des prêts à l'investissement (par opposition aux prêts d'ajustement) dans l'agriculture d'un montant de 2,8 milliards de dollars, soit 8 % du montant total des prêts à l'investissement qu'elle a accordés à la région. Une part substantielle de ces prêts a revêtu la forme de composantes agricoles de projets ruraux. La Banque a par ailleurs approuvé 77 prêts à l'appui des politiques de développement comportant un volet agricole. Dans 18 d'entre eux, l'agriculture était une composante appréciable.

Ces prêts limités à l'investissement ont obtenu des résultats insuffisants. Selon les données du GIE, le pourcentage de projets d'investissement essentiellement agricoles dont les résultats ont été jugés satisfaisants entre 1991 et 2006 (60 %) est inférieur à celui des projets d'investissement non agricoles dans la région (65 %). Il est aussi inférieur à celui de projets d'investissements similaires dans d'autres régions de la Banque mondiale, dont 73 % ont affiché des résultats satisfaisants. Leurs notes de viabilité ont également été inférieures à la moyenne. Bien que d'autres analyses s'imposent, l'étude a observé que les projets à forte composante agricole conduits dans les pays où les conditions sont moins propices à l'agriculture ont enregistré de meilleurs résultats que des projets analogues exécutés dans des pays où les conditions sont plus favorables.

Pour appuyer le développement agricole en Afrique, la Banque mondiale a axé ses opérations sur les prêts, les études analytiques et les activités de conseil. Jusque très récemment, les études analytiques – nécessaires au diagnostic des problèmes et à la définition d'interventions, ainsi qu'à la formulation de conseils stratégiques et aux opérations de prêt – étaient rares, dispersées, de qualité variable et peu accessibles. L'IEG a en outre constaté qu'il n'existait pas de procédures particulières pour vérifier que les dialogues avec les autorités et les négociations de prêts prennent systématiquement en compte les conclusions de ces études.

L'IEG a observé que les prêts accordés par la Banque mondiale ne tenaient pas compte de l'interdépendance des activités agricoles. Au contraire, ils ont été répartis sur un éventail d'opérations portant sur le secteur rural, notamment la recherche, la vulgarisation, la réforme de la commercialisation, la lutte contre la sécheresse, le développement de semences et les transports, mais ont fait peu de cas des liens qui les unissent et de la nécessité de développer ces domaines simultanément, ou tout du moins selon un enchaînement optimal, pour favoriser efficacement le développement agricole. Si, à compter du milieu des années 80, la stratégie rurale plus globale de la Banque se justifiait, elle a eu pour conséquence imprévue de faire perdre de vue la nécessité de conduire certaines activités essentielles au développement de l'agriculture dans l'espace rural simultanément ou selon un agencement optimal.

L'examen de l'IEG a constaté qu'aucun des dix premiers pays emprunteurs, dont la Côte d'Ivoire, l'Éthiopie, l'Ouganda et la Tanzanie, n'avait bénéficié d'une aide *homogène* et *simultanée* dans tous les sous-secteurs critiques. Il ne s'agit pas de proposer que la Banque mondiale apporte seule ce type d'assistance (un partenariat serait selon toute probabilité plus efficace), mais on pourrait s'attendre à ce qu'elle prenne l'initiative d'encourager une démarche pluridimensionnelle de cette nature, fondée sur son avantage comparatif en tant qu'institution de prêt multisectorielle.

Les résultats par domaines

Une évaluation des réussites et des lacunes de l'aide apportée par la Banque mondiale dans chaque grand domaine dévoile un bilan mitigé :

Diversité agroécologique. L'assistance de la Banque mondiale a permis de renforcer la capacité des systèmes nationaux de recherche et de mettre sur pied des stations dans les différentes zones pour orienter les études sur l'agroécologie. Il ne semble cependant pas que ses projets dans des domaines autres que la recherche aient systématiquement adapté les opérations aux spécificités agroécologiques variées des différents pays. L'attrait essentiel des projets qui font appel à des stratégies communautaires était leur aptitude à s'adapter aux situations locales, mais rien ne permet d'affirmer que ces stratégies, telles qu'appliquées dans les projets conduits au Ghana et en Tanzanie, par exemple, soient capables de tenir compte de la diversité agroécologique.

Variabilité des précipitations et des sécheresses. Les projets que la Banque mondiale avaient mené à terme à la fin de l'exercice 06 ont tenu compte des situations d'urgence créées par les sécheresses, et ont aidé les autorités à établir des dispositifs de gestion des sécheresses. Ils n'ont cependant pas été en mesure d'aider des pays tels que le Malawi, par exemple, à élaborer une approche stratégique de long terme pour remédier aux causes fondamentales de l'insécurité alimentaire – autrement dit, à aider les pays à augmenter suffisamment leur productivité agricole afin de mettre un terme à la diminution des ressources alimentaires par habitant. À cet égard, si la Banque a concouru au développement de variétés améliorées de millet et de manioc au travers de son aide à la recherche, elle n'a pas su mesurer le rôle capital que le manioc peut jouer au service de la sécurité alimentaire dans la majorité des pays.

Faible fertilité des sols. La Banque mondiale a participé à plusieurs projets internationaux et régionaux dans ce domaine, notamment à l'initiative régionale *TerrAfrica,* lancée en 2005. Ce partenariat pluridimensionnel est censé favoriser une approche collective à la gestion durable des terres dans la région. Les opérations de prêt de la Banque semblent toutefois avoir abordé la fertilité des sols sous un angle écologique plutôt qu'en termes de productivité agricole.

Approvisionnement en eau. Bien que la Banque mondiale ait établi que des investissements dans l'irrigation s'imposaient, elle n'a consacré que très peu de prêts à cet objectif. Ses interventions à l'appui de la gestion de l'eau dans les zones non irriguées ont atteint des objectifs matériels mais, étant donné la médiocrité du suivi et de l'évaluation, il est difficile de définir ce qui a fonctionné ou pas.

Semences améliorées. La Banque mondiale a participé au Groupe consultatif pour la recherche agricole internationale (CGIAR), dont le concours dans ce domaine a été notable. Les projets qu'elle a conduits ont également offert la possibilité de tester les technologies et de les développer, comme en Éthiopie et au Togo. Cela étant, les opérations associées aux semences n'ont à ce stade apporté qu'une modeste contribution à l'augmentation de la production agricole. Les projets de la Banque mondiale n'ont pas non plus été en mesure de résoudre le problème de l'usage limité que les agriculteurs font des semences en raison de l'accès difficile aux intrants complémentaires.

Accès des agriculteurs au crédit et à la finance rurale. Globalement, l'aide de la Banque mondiale dans ce domaine stratégique a été limitée. Outre l'insuffisance des capacités institutionnelles des pays clients, cette faiblesse s'explique par les mauvais résultats des projets dans ce domaine qui tiennent, entre autres, à l'application déficiente des directives de la Banque, surtout en ce qui concerne les critères de sélection des intermédiaires financiers et de leurs prestations. La Banque mondiale doit élaborer et superviser plus soigneusement ces opérations, et continuer d'examiner toutes les solutions envisageables pour définir au mieux comment donner aux agriculteurs les moyens dont ils ont besoin pour accroître leur productivité et leurs revenus.

Médiocrité de l'infrastructure de transports. Les interventions financées par la Banque mondiale dans le secteur agricole n'ont que faiblement concouru à l'amélioration de l'infrastructure de transport pour faciliter l'accès aux marchés et, partant, le développement de l'agriculture.

Insuffisance des opérations de vulgarisation agricole. La Banque mondiale a sensibilisé les clients à l'importance de la vulgarisation pour le développement agricole. Elle appuie actuellement divers projets conduits en partenariat (partenariats public-privé ou déterminés par la demande, organisations non gouvernementales, etc.), en Ouganda par exemple. Il faudrait toutefois systématiquement évaluer les coûts, l'efficacité et la viabilité de ces projets.

Réforme des prix et de la commercialisation. Bien que les résultats varient selon les pays, les mesures appliquées par la Banque mondiale ont permis d'améliorer la conjoncture macroéconomique et la discipline budgétaire dans plusieurs pays. Ces réformes n'ont cependant pas suffi à stimuler l'investissement privé dans divers domaines stratégiques d'où le secteur privé s'est retiré. Par conséquent, la plupart des pays africains font face aux prix exorbitants des engrais, à la production insuffisante de semences, à un système de transports défectueux, et à l'accès limité au crédit. Si le processus de réformes n'a eu que peu de retombées favorables sur la production alimentaire, il a dynamisé la production de cultures d'exportation non traditionnelles, comme les mangues au Mali et les fleurs au Kenya. Au-delà de son action à l'échelle de chaque pays, la Banque est intervenue pour que le cycle de négociations de Doha serve véritablement le développement et a plaidé en faveur de la suppression des subventions à l'agriculture des pays de l'OCDE dans les instances internationales. Elle n'a toutefois obtenu que peu de résultats à ce jour.

Précarité du régime foncier. Les études analytiques réalisées ont permis de mieux appréhender les régimes de droits fonciers. Étant donné ses sensibilités politiques, sociales et culturelles, la Banque mondiale a cependant jugé difficile d'apporter une aide efficace dans ce domaine.

En 2005, le Groupe de travail sur la faim du projet Objectifs du Millénaire a abouti à la conclusion que l'ODM consistant à diminuer la faim de moitié d'ici à 2015 était réalisable. Le développement de l'agriculture africaine est l'élément capital à la concrétisation de cet objectif. La Banque mondiale est en mesure d'y apporter un concours essentiel, puisqu'elle est l'une des sources majeures de financement du développement pour l'agriculture et peut conseiller les autorités quant aux mesures à adopter.

Conclusions essentielles quant aux facteurs de performance de la Banque mondiale et des pays

Les facteurs de performance de la Banque mondiale

- La stratégie de l'institution pour le développement du secteur agricole s'inscrit dans le cadre de sa stratégie rurale et, avec le temps, la place de l'agriculture dans cette stratégie a rétréci. En conséquence, les compétences techniques nécessaires pour œuvrer valablement au développement agricole se sont peu à peu émoussées, phénomène qui a à son tour entretenu cette situation. Selon les chiffres communiqués par le Département des ressources humaines de la Banque mondiale, 17 experts techniques du Département de l'agriculture et du développement rural étaient chargés de l'Afrique subsaharienne en 2006, alors qu'ils étaient 40 en 1997.
- C'est essentiellement au travers d'études analytiques que la Banque définit l'état de développement d'un pays et les priorités dans le secteur agricole. Jusque très récemment, ces études étaient peu nombreuses et difficilement accessibles. Qui plus est, leurs conclusions n'ont guidé ni le dialogue de la Banque avec les autorités des pays clients, ni l'élaboration des programmes de prêt.
- Les conseils de la Banque semblent avoir fortement influencé l'orientation du développement agricole dans les pays africains, notamment ceux associés au programme d'ajustement. Les résultats n'ont toutefois pas été à la hauteur des attentes en raison, notamment, d'un appui politique insuffisant et d'une appréciation incomplète des réalités sur le terrain.
- Les systèmes de données de la Banque mondiale et l'aide au suivi et à l'évaluation n'ont pas permis d'éclairer de manière satisfaisante les opérations engagées par l'institution pour développer l'agriculture sur un large front en Afrique. Les systèmes de données actuels ne lui permettent pas de suivre suffisamment en détail les montants alloués à la mise en place d'opérations spécifiques comme le dévelop-

pement des semences et le crédit. Le suivi et l'évaluation à l'échelon des projets n'ont guère permis de répondre aux questions fondamentales portant sur leurs résultats, leurs retombées et leur efficacité, pour savoir par exemple quels en ont été les bénéficiaires, quelles cultures ont bénéficié d'une aide et selon quelles modalités, quelle a été leur rentabilité comparative, et à quels facteurs les progrès sont imputables.

Les facteurs nationaux

- Malgré les failles persistantes de la gouvernance dans plusieurs pays africains, l'engagement politique des pays clients envers le développement de l'agriculture semble plus ferme qu'auparavant. Les gouvernements africains, dont beaucoup allouaient moins de 1 % de leur budget à l'agriculture, ont convenu au Sommet de l'Union africaine, qui s'est tenu en juillet 2003, d'affecter au moins 10 % des ressources budgétaires nationales à des programmes destinés à favoriser la croissance agricole au cours des cinq prochaines années.
- La capacité de recherche agronomique existante est considérable, mais la viabilité des opérations soutenues par la Banque reste aléatoire. Dans l'ensemble, la capacité de l'administration publique de plusieurs pays demeure insuffisante, et les ministères de l'agriculture sont encore des partenaires relativement inefficaces pour favoriser le développement du secteur agricole. Bien que d'autres analyses s'imposent, l'étude a révélé que les projets à forte composante agricole exécutés dans les pays où les conditions sont moins propices à l'agriculture ont enregistré de meilleurs résultats que des projets analogues conduits dans des pays aux conditions plus favorables, ce qui semble indiquer que d'autres facteurs, tels que l'économie politique et la capacité nationale, font également obstacle au développement de l'agriculture en Afrique.

Recommandations

Pour appuyer utilement la mise en œuvre du Plan d'action pour l'Afrique et l'objectif prioritaire de développement agricole qu'il s'est judicieusement fixé, l'IEG formule à l'intention de la Banque mondiale les recommandations suivantes :

1. Privilégier l'amélioration de la productivité agricole :
   - Établir des objectifs réalistes en matière d'expansion de l'irrigation et prendre conscience de la nécessité d'augmenter la productivité de l'agriculture non irriguée par l'amélioration de la qualité des terres et par la gestion de l'eau et des sécheresses.
   - Participer à l'élaboration de mécanismes efficients, dont des partenariats public-privé, pour fournir aux agriculteurs les intrants indispensables, notamment des engrais, de l'eau, des crédits et des semences.
   - Appuyer le développement des infrastructures de commercialisation et de transport.
2. Améliorer ses études dans le domaine de l'agriculture :
   - Augmenter la quantité et la qualité des études analytiques portant sur l'agriculture et veiller à ce que ses opérations de conseil et de prêt se fondent sur leurs conclusions.
   - Appuyer les analyses des dépenses publiques de manière à évaluer les ressources disponibles pour l'agriculture et à définir ses propres priorités.
   - Restaurer ses compétences techniques à partir d'une évaluation complète de ses lacunes actuelles.
3. Établir des indicateurs afin de mesurer les progrès :
   - Perfectionner les systèmes de données pour mieux suivre les opérations qu'elle finance.
   - Consolider le dispositif de suivi et d'évaluation pour rendre compte des activités conduites dans le cadre des projets dans diverses zones agroécologiques, et pour différentes cultures et catégories d'agriculteurs, y compris les femmes.
   - Mettre sur pied un système permettant de coordonner les opérations agricoles à l'échelle des pays en fonction de leur réseau routier, de la proximité des marchés et des conditions édaphiques.

Resumo Executivo

A África Subsariana é uma Região muito complexa, composta por 47 países com 7 histórias coloniais nitidamente diferentes. É também uma Região muito diversa, com mais de 700 milhões de pessoas e pelo menos 1.000 grupos étnicos diferentes. A Região constitui uma prioridade principal para o desenvolvimento. Nela encontram-se alguns dos países mais pobres do mundo, e durante as duas últimas décadas o número de pessoas pobres na Região duplicou, passando para 300 milhões— mais de 40 por cento da população da Região. A África continua a ficar para trás no que respeita a maioria dos Objectivos de Desenvolvimento para o Milénio (ODM) e é pouco provável que os alcance até ao ano de 2015.

O principal impedimento ao desenvolvimento da África é o fraco desempenho do sector da agricultura. Este sector reveste uma importância fundamental para a Região visto que tem uma grande participação no produto interno bruto (PIB) e no emprego. O fraco desempenho do sector deriva de vários constrangimentos que são específicos à agricultura em África e tornam o seu desenvolvimento um desafio complexo. A debilidade da governação e os conflitos em diversos países complicam as coisas ainda mais. O IEG avaliou a eficácia para o desenvolvimento da assistência prestada pelo Banco Mundial ao abordar os constrangimentos existentes para o desenvolvimento da agricultura na África durante o período compreendido entre os anos fiscais de 1991 e 2006, a título experimental, para depois realizar uma avaliação mais ampla da assistência prestada pelo Banco à agricultura em todo o mundo.

A conclusão central do estudo é que o sector da agricultura tem sido negligenciado tanto pelos governos como pela comunidade de doadores, incluindo o Banco Mundial. A estratégia do Banco para a agricultura está cada vez mais subordinada ao âmbito de uma focalização rural mais ampla, o que diminuiu a sua importância. Decorrente disso e contribuindo para isso, a competência técnica para apoiar adequadamente o desenvolvimento da agricultura também diminuiu com o tempo.

O apoio limitado, e até recentemente decrescente, do Banco para abordar os constrangimentos à agricultura não foi utilizado estrategicamente para suprir as necessidades diversas de um sector que requer uma intervenção coordenada numa grande variedade de actividades. O apoio dos empréstimos do Banco tem sido "espalhado" por várias actividades agrícolas, tais como, investigação, extensão, crédito, sementes e reformas de política no espaço rural, mas a sinergia potencial entre eles, para contribuir efectivamente para o desenvolvimento da agricultura, pouco foi reconhecida. Consequentemente, o Banco tem tido um êxito limitado na sua contribuição para o desenvolvimento da agricultura na África.

Os Desafios da Agricultura Africana

A produção agrícola aumentou na África, mas é difícil calcular uma taxa de crescimento confiável para a Região durante o período em estudo devido às grandes variações que existem entre os países e através do tempo. Alguns países, tais como o Gabão, passaram de um desempenho fraco em 1990 – 2000 para um desempenho melhor em 2000 – 2004; outros países, tais como o Malawi, mudou num sentido diferente. A modificação foi frequentemente espectacular, o que faz com que as taxas de crescimento agregadas sejam enganosas. Por exemplo, a agricultura em Angola cresceu a um ritmo de 13,7 por cento anuais durante o período de 2000 – 2004, embora tenha havido uma contracção do crescimento de 1,4 por cento anuais durante o período de 1990-2000. Apenas um quarto dos países da Região, entre os quais o Benin, o Burkina Faso, o Gana, a Nigéria e a Tanzânia mostram um crescimento constante da agricultura de mais de 3 por cento no período de 1990 – 2004.

A produção agrícola total da África consiste principalmente em cultivos alimentares. Os produtos agrícolas para exportação representam menos de 10 por cento da produção total. Se bem que alguns produtos de exportação, incluindo o algodão, tenham contribuído para aliviar a pobreza em países como o Burkina Faso, os cultivos alimentares tiveram um fraco desempenho na maioria dos países. A produção de cereais em África, mesmo em 2003— 2005, foi menos de metade da do Sul da Ásia e um terço da produção da América Latina. A África também está atrás de outras regiões na percentagem de terrenos cultivados irrigados, na utilização de fertilizantes, e na produtividade dos trabalhadores e das terras. Embora os grandes avanços na produção agrícola da Ásia do Sul entre 1961 e 2001 tenham sido principalmente o resultado de maiores rendimentos, os aumentos da produção de alimentos em África foram produzidos sobretudo através da ampliação das terras cultivadas. Entretanto, o rendimento dos cultivos estagnou.

A partir de 1973 a África tornou-se um importador de bens alimentares. Desde essa época a produção de alimentos não acompanha o ritmo de crescimento rápido da população e as importações de alimentos aumentaram com celeridade. Entretanto, as exportações da África que se baseiam principalmente na agricultura, decresceram; no caso de certos produtos, incluindo o café, a participação da Região no mercado mundial evaporou-se. Os subsídios à agricultura nos países membros da Organização para a Cooperação e o Desenvolvimento Económico (OCDE) tiveram um importante papel em manter baixos os preços mundiais de vários desses produtos agrícolas. Isto, entre outros factores, teve repercussões sobre a adequação dos rendimentos dos agricultores.

A agricultura na África é principalmente uma actividade familiar, e a maioria dos agricultores são pequenos proprietários de terrenos com uma extensão que varia entre 0,5 e 2,0 hectares, conforme determinarem os factores sócio-culturais. As mulheres fornecem cerca de metade da força de trabalho e produzem a maioria dos produtos alimentares que a família consome.

As terras agrícolas em África encontram-se em várias zonas agro-ecológicas que se estendem por diversos países. Caracterizam-se em grande medida por solos de má qualidade, pluviosidade muito variável, e secas frequentes. As infra-estruturas de transportes são deficientes, o acesso à irrigação é limitado e, nas condições de cultivos de sequeiro, a insegurança alimentar é uma realidade para milhões de pequenos agricultores. Para sobreviverem neste ambiente inauspício, a maior parte dos agricultores seguem uma diversidade de estratégias para enfrentarem a situação. Para assegurar que obtêm pelo menos alguns produtos agrícolas das suas terras, os agricultores africanos plantam normalmente uma variedade de cultivos (normalmente 10 ou mais) com diferentes períodos de maturação, juntamente com árvores. O gado também é uma fonte importante de segurança para os agricultores da África, especialmente em anos de escassez. O acesso a crédito pelo pequeno proprietário médio também é extremamente limitado. Os cultivos resistentes, tais como o milho-miúdo, o sorgo, a mandioca e outras plantas tuberosas são mais importantes do que os cereais, como o arroz e o trigo, que foram o alimento principal da Revolução Verde da Ásia.

Nesta situação, para que os agricultores tenham um incentivo para praticarem a agricultura intensiva e assumam riscos com novas variedades de cultivos, é necessária a conjugação de alguns factores, ou pelo menos que eles se apresentem numa sequência óptima, o que inclui sementes melhoradas, água, crédito e acesso aos mercados; boa assessoria sobre os serviços de extensão; e rendimentos adequados por meio de preços dos insumos e produtos isentos de distorções. Uma estratégia para o desenvolvimento da agricultura na África tem que levar em conta cada um destes factores no contexto das suas características singulares e das condições locais específicas da África.

As Abordagens do Passado à Agricultura Africana

Até muito recentemente, o desenvolvimento da agricultura na África foi negligenciado, tanto pelos governos como pelos doadores. Durante a década de 1960, imediatamente após a independência, os governos de vários países africanos consideravam que a agricultura era principalmente fonte de recursos para a industrialização. Em seguida, na década de 1970, o Banco Mundial liderou a mudança para um modelo de desenvolvimento mais amplo da África que era compatível com uma mudança mais geral na compreensão do desenvolvimento. Com isto, a instituição comprometeu-se a seguir a via do desenvolvimento rural integrado, destinada a atacar directamente a pobreza e o subdesenvolvimento rurais na África. Em meados da década de 1980, quando os países africanos enfrentavam crises fiscais graves, os doadores deram a prioridade a melhorar a eficiência da repartição de recursos e pressionaram no sentido de serem adoptadas reformas da comercialização dos produtos agrícolas. Mas as reformas estruturais também não chegaram a produzir os efeitos desejados sobre o crescimento.

O Papel das Ajudas

A assistência bilateral e multilateral dos doadores para o desenvolvimento da agricultura na África diminuiu, passando de USD 1.921 milhões em 1981 para USD 997 milhões em 2001 (em dólares de 2001). Os empréstimos de ambas as fontes ressurgiram desde então, com uma focalização crescente no desenvolvimento da África. Os dados da OCDE mostram que, se bem que os doadores bilaterais agrupados tenham tido um papel comparativamente mais importante, o Banco Mundial por si só foi o maior doador para a agricultura africana entre 1990 e 2005. Os maiores doadores bilaterais foram os Estados Unidos e o Japão.

Os fluxos de capital estrangeiro privado para a África são modestos em comparação com as ajudas bilaterais e multilaterais (Hazell e von Braun 2006). O investimento privado comercial na agricultura africana tem-se limitado em grande medida aos produtos agrícolas para exportação e às zonas com um maior potencial. Algumas empresas de sementes internacionais investiram na multiplicação da semente do milho, e em Setembro de 2006 as fundações de Rockefeller e de Bill e Melinda Gates iniciaram conjuntamente uma nova parceria para ajudar a África a desenvolver a agricultura.

O Potencial da Agricultura e a Estratégia do Banco

Para que a África alcance os Objectivos de Desenvolvimento para o Milénio (ODM) será necessário realizar o potencial do sector da agricultura, para prestar o apoio a que contribua para o crescimento e a redução da pobreza. A investigação realizada por Dorosh e Haggblade (em 2003) e o IFPRI (2006a) chegou à conclusão que os investimentos na agricultura favorecem geralmente as populações pobres da África mais do que investimentos semelhantes na indústria manufactureira.

O Banco Mundial não tem tido uma estratégia distinta para a agricultura na África, excepto como fazendo parte das suas estratégias mais amplas para o desenvolvimento rural, e com o tempo a estratégia para a agricultura foi subordinada a uma focalização mais ampla nas zonas rurais. Porém, mais recentemente, o Plano de Acção para a África reconheceu que o sector da agricultura era potencialmente o motor do crescimento.

A Assistência do Banco em Termos Globais e a Avaliação da Mesma

Durante os anos fiscais compreendidos entre 1991 e 2006, o Banco forneceu aos países da Região da África USD 2,8 biliões sob a forma de empréstimos para o desenvolvimento (o que é distinto dos empréstimos para o ajustamento) da agricultura, o que constituiu 8 por cento de todos os empréstimos do Banco à Região para o desenvolvimento. Uma grande parte destes empréstimos foi concedida sob a forma de componentes da agricultura nos projectos rurais. Adicionalmente, foram concedidos 77 Empréstimos para a Política de Desenvolvimento com componentes para a agricultura, e em 18 deles a agricultura teve uma dimensão importante.

Estes empréstimos limitados para o desenvolvimento tiveram um desempenho abaixo do normal. Os dados do IEG mostram que a percentagem de classificações de resultados satisfatórios para projectos de investimento sobretudo na agricultura durante o período de 1991 – 2006 é inferior à dos investimentos não relacionados com a agricultura na Região (60 por cento face a 65 por cento satisfatórios). Também é inferior a percentagem para os projectos de investimento semelhantes noutras Regiões do Banco (73 por cento satisfatórios). As classificações da sustentabilidade também são inferiores à média. Embora seja necessário realizar análises adicionais, o estudo verificou que os projectos que se destinam em grande medida à agricultura em países com condições menos favoráveis para a agricultura tiveram melhores resultados do que os projectos semelhantes em países com condições mais favoráveis.

As actividades do Banco de apoio ao desenvolvimento da agricultura na África compreenderam empréstimos, trabalho analítico e assessoria política. Até muito recentemente o trabalho analítico, que é necessário para fazer o diagnóstico das questões e acções e para ajudar a definir a assessoria política e os empréstimos, foi limitado, disperso, de qualidade variável e não era facilmente disponível. Adicionalmente, o IEG verificou que não existem procedimentos específicos para assegurar que as conclusões do trabalho analítico sejam reflectidas sistematicamente nos empréstimos e no diálogo sobre as medidas de política.

O IEG verificou que o apoio prestado pelo Banco sob a forma de empréstimos não reflectiu a interligação existente entre as várias actividades na agricultura. Em vez disso, os empréstimos foram "espalhados" por uma variedade de actividades no espaço rural, incluindo a investigação, a extensão, as reformas da comercialização, o alívio às secas, o desenvolvimento de sementes e os transportes, mas pouco com um escasso reconhecimento das relações entre elas e da necessidade de desenvolver todas estas áreas simultaneamente ou pelo menos numa sequência óptima, para contribuir efectivamente para o desenvolvimento da agricultura. Embora a focalização mais ampla do Banco nas zonas rurais a partir de meados de 1980 fosse justificada, um resultado indesejado foi ela levou a prestar menos a atenção à necessidade de conjugar as várias actividades que são cruciais para o desenvolvimento da agricultura no espaço rural ou se elas terem lugar numa sequência óptima.

Esta análise verificou que nenhum dos dez maiores mutuários, entre os quais a Côte d'Ivoire, a Etiópia, a Tanzânia, e o Uganda, receberam um apoio *consistente* e *simultâneo* em todos os subsectores fundamentais. Isto não é uma sugestão para que o Banco o faça sozinho— poderá mesmo ser feito melhor em parceria— mas poder-se-ia esperar razoavelmente que o Banco assuma a liderança na promoção desta abordagem multifacetada, com base na sua vantagem comparativa como instituição que concede empréstimos multissectoriais.

O Desempenho Temático

A avaliação das realizações e limitações do apoio prestado pelo Banco por tema principal revela uma experiência mista:

A diversidade agro-ecológica. O apoio do Banco ajudou a aumentar as capacidades dos sistemas nacionais de investigação e a desenvolver estações zonais para dar uma focalização agro-ecológica à investigação. Contudo, pouco indica que os projectos do Banco, a não ser as intervenções de investigação, adaptaram sistematicamente as actividades às condições agro-ecológicas diversas. A capacidade de resposta às condições locais tem sido o atractivo principal dos projectos que seguem abordagens baseadas nas comunidades, mas há poucas provas de que estas abordagens, tal como foram seguidas em projectos no Gana e na Tanzânia, por exemplo, possam dar resposta à diversidade agro-ecológica.

A flutuação da pluviosidade e das secas. Os projectos do Banco concluídos até ao fim do ano fiscal de 2006 foram responsivos às emergências causadas pelas secas e ajudaram os governos a instaurar sistemas de gestão das secas. Mas não conseguiram ajudar países, como o Malawi por exemplo, a desenvolver uma estratégica a longo prazo para abordar os factores básicos que criam a insegurança alimentar, isto é, para ajudar os países a aumentarem a produtividade agrícola de maneira suficiente para travar o declínio da disponibilidade per capita de alimentos. A este respeito, se bem que o Banco tenha contribuído para o desenvolvimento de variedades melhoradas de milho-miúdo e de mandioca, mediante o apoio prestado à investigação, ele perdeu a oportunidade de reconhecer o papel importante que a mandioca pode ter na promoção da segurança alimentar na maioria dos países.

Os solos pouco férteis. O Banco participou em diversas iniciativas internacionais e regionais sobre este assunto, incluindo a *Iniciativa Regional Terr África* teve início em 2005. Esta parceria multidimensional deverá promover uma abordagem colectiva à gestão sustentável das terras na Região. Mas parece que os empréstimos do Banco abordaram a fertilidade dos solos mais como uma questão ambiental do que uma questão de produtividade agrícola.

O acesso a água. Embora o Banco tenha identificado a necessidade de investimentos na irrigação, os empréstimos que concedeu para esse fim foram muito limitados. As intervenções do Banco que apoiam a gestão dos recursos hídricos nas zonas de sequeiro atingiram as metas físicas mas, devido à debilidade da monitorização e avaliação, é difícil dizer o que funcionou e não funcionou.

As sementes melhoradas. O Banco contribuiu para o Grupo Consultivo sobre a Investigação Agrícola Internacional (CGIAR), fazendo grandes qual contribuições nesta área, e os projectos do Banco também proporcionaram a oportunidade de testar e melhorar as tecnologias, como na Etiópia e no Togo. No entanto, as actividades relacionadas com as sementes fizeram até agora apenas uma contribuição modesta para o aumento da produção dos cultivos. Os projectos do Banco também não puderam abordar a questão da utilização limitada de sementes pelos agricultores devido a um acesso inadequado a insumos complementares.

O acesso dos agricultores ao crédito e ao financiamento rural. Em termos globais, o apoio prestado pelo Banco nesta área crucial foi limitado. Para além da debilidade das capacidades institucionais nos países clientes, uma razão para este baixo nível de apoio foi o fraco desempenho dos projectos nesta área, provocado, entre outras coisas, pela aplicação deficiente das directrizes do Banco, em particular no que respeita a elegibilidade e o desempenho dos intermediários financeiros. É necessário que o Banco tenha mais cuidado na concepção e supervisão destas actividades, devendo continuar a ser exploradas todas as opções para dar aos agricultores de maneira mais apropriada os meios necessários para aumentarem a produtividade e os rendimentos.

A debilidade das infra-estruturas de transportes. As intervenções na agricultura apoiadas pelo Banco apenas fizeram uma contribuição limitada ao melhoramento das infra-estruturas de transportes para o desenvolvimento da agricultura.

A debilidade da extensão. O Banco auxiliou os seus clientes a tomarem consciência da importância da extensão para o desenvolvimento da agricultura. Actualmente o Banco apoia várias abordagens de parcerias (públicas-privadas, motivadas pela procura, organizações não governamentais, etc.), como no Uganda. Mas é preciso avaliar sistematicamente o custo, a eficácia e a sustentabilidade dessas abordagens.

A reforma dos preços e da comercialização. Embora os resultados tenham variado de um país para o outro, o esforço do Banco contribuir para melhorar a conjuntura macroeconómica e a disciplina fiscal em vários países. Contudo, essas mudanças não foram suficientes para estimular os investimentos do sector privado em diversas áreas cruciais das quais o sector público se retirou. Por conseguinte, a maioria dos países africanos vêm-se perante preços exorbitantes dos fertilizantes, uma produção inadequada de sementes, transportes deficientes, e um acesso limitado ao crédito. Se bem que o processo de reformas tenha tido um efeito positivo limitado sobre a produção de alimentos, ele no entanto aumentou a produção de produtos alimentares de exportação não tradicionais, como as mangas do Mali e as flores do Quénia. Para além dos países individuais, o Banco exerceu pressão para que a ronda de negociações de Doha fosse autenticamente a favor do desenvolvimento, e para a eliminação dos subsídios à agricultura nos países da OCDE nos fóruns internacionais, mas até à data com pouco êxito.

A insegurança da propriedade das terras. O trabalho analítico contribuiu para uma melhor compreensão dos regimes de direitos de propriedade. Mas o Banco teve dificuldade em prestar um apoio efectivo nesta área devido às sensibilidades políticas, sociais e culturais.

O Grupo de Trabalho sobre a Fome do Projecto de Desenvolvimento do Milénio concluiu em 2005 que o mundo poderia alcançar os Objectivos de Desenvolvimento para o Milénio (ODM) se reduzisse a fome para metade até ao ano de 2015. O desenvolvimento da agricultura da África é crucial para alcançar este objectivo e o Banco Mundial pode fazer uma contribuição importante porque é uma das maiores fontes de financiamento do desenvolvimento para a agricultura e pode prestar assessoria política aos governos.

Principais Conclusões sobre os Factores Relacionados com o Desempenho do Banco e dos Países

Factores relacionados com o Banco

- A estratégia da instituição para o desenvolvimento do sector da agricultura tem feito parte da sua estratégia rural, e com o tempo a importância da agricultura na estratégia rural do Banco diminuiu. Como resultado disso e contribuindo para isso, as competências técnicas para apoiar o desenvolvimento da agricultura também diminuíram com o tempo. Os dados proporcionados pelo Departamento de Recursos Humanos do Banco Mundial mostram que havia 17 especialistas técnicos destacados ao Departamento de Desenvolvimento da Agricultura e Rural na África Subsariana em 2006, comparado com 40 em 1997.
- O diagnóstico do Banco da situação de desenvolvimento e das prioridades do sector da agricultura dum país é realizado principalmente através do trabalho analítico. Até muito recentemente, este trabalho era limitado e não era facilmente disponível. As conclusões do trabalho analítico também não informaram estrategicamente o diálogo de política com os clientes do Banco nem a concepção dos programas para os empréstimos.
- A política do Banco parece ter tido implicações de longo alcance para a direcção do desenvolvimento da agricultura nos países africanos, em especial a sua assessoria de política associada à agenda do ajustamento. Contudo, os resultados foram inferiores às expectativas devido, entre outras coisas, a um apoio político débil e a um conhecimento insuficiente da realidade no terreno .
- Os sistemas de dados do Banco e o apoio à monitorização e avaliação foram insuficientes para informar adequadamente os esforços da instituição para desenvolver a agricultura na África num âmbito amplo. Os sistemas de dados actuais não permitem que a instituição faça um seguimento suficientemente detalhado dos montantes proporcionados para realizar actividades específicas, tais como o desenvolvimento de sementes e o crédito. O monitoramento e a avaliação a nível dos pro-

jectos tem tido um valor limitado para responder às questões fundamentais sobre os resultados, as repercussões e a eficiência, tais como, quem beneficia, que cultivos receberam apoio e de que maneira, qual tem sido a eficácia comparativa dos custos e a que podem ser atribuídos os ganhos.

Factores relacionados com o país

- Embora a conjuntura da governação em vários países africanos continue a ser débil, o comprometimento politico para o desenvolvimento da agricultura nos países clientes parece ser mais forte do que no passado. Os governos africanos, muitos dos quais afectavam menos de 1 por cento dos seus orçamentos à agricultura, concordaram em Julho de 2003 na Cimeira da União Africana em afectar pelo menos 10 por cento das verbas do orçamento nacional a programas destinados a apoiar o crescimento da agricultura nos próximos cinco anos.
- Existem capacidades consideráveis para a investigação na agricultura, se bem que a sustentabilidade das actividades que recebem apoio continue a ser incerta. Em termos globais, as capacidades do governo em vários países continuam a ser débeis, e os ministérios da agricultura nacionais ainda são parceiros relativamente ineficazes na promoção do desenvolvimento do sector da agricultura. Embora seja necessário realizar análises adicionais, a conclusão do estudo de que os projectos destinados em grande medida à agricultura em países com condições agrícolas menos favoráveis produziram melhores resultados do que os projectos semelhantes em países com condições mais favoráveis, sugere que outros factores, tais como a economia política e as capacidades do país, também constituem um desafio para o desenvolvimento da agricultura na África.

Recomendações

Para apoiar efectivamente a execução do Plano de Acção para a África e a sua focalização apropriada no desenvolvimento da agricultura como sendo uma prioridade fundamental, o IEG recomenda que o Banco:

1. Concentre a sua atenção na consecução de melhoramentos da produtividade agrícola.
   - Defina objectivos realistas para a ampliação da irrigação e reconheça a necessidade de aumentar a produtividade da agricultura de sequeiro mediante o melhoramento da qualidade das terras, assim como a gestão dos recursos hídricos e das secas.
   - Ajude a conceber mecanismos eficientes, incluindo parcerias entre os sectores público e privado, para fornecer aos agricultores os insumos fundamentais, incluindo fertilizantes, água, crédito e sementes.
   - Preste apoio ao desenvolvimento das infraestruturas de comercialização e de transportes.
2. Melhore o trabalho que realiza na área da agricultura:
   - Aumente a quantidade e a qualidade do trabalho analítico sobre a agricultura e assegure que a assessorial política e os empréstimos se baseiem nas conclusões desse trabalho.
   - Apoie as análises das despesas públicas para avaliar a disponibilidade de recursos para a agricultura e ajudar a definir as prioridades do Banco.
   - Aumente novamente as suas competências técnicas, com base numa avaliação abrangente das lacunas actuais.
3. Estabeleça referências para medir os progressos realizados:

- Aperfeiçoe os sistemas de dados para melhor fazer o seguimento das actividades que recebem apoio do Banco.
- Reforce o monitoramento e a avaliação para informarem sobre as actividades dos projectos nas diversas zonas agro-ecológicas e para as categorias diferentes de cultivos e agricultores, incluindo as mulheres.
- Desenvolva um sistema para coordenar as actividades agrícolas num país com o acesso rodoviário, a proximidade dos mercados e as condições dos solos.

Cultivating a field in Senegal. Photo by Ray Witlin, courtesy of World Bank Photo Library.

Management Response

Management agrees with and is already implementing the main message of this review—that is, that the World Bank should provide more support for agriculture in Sub-Saharan Africa. Management has comments on several elements of IEG's analysis and differs with some of IEG's views on the way forward to achieve the shared objectives. The main points are:

- Management agrees that to achieve the Millennium Development Goals (MDGs) in Sub-Saharan Africa, agriculture must play a more effective role. The Region, in close collaboration with African leaders and development partners, has already moved the Bank significantly in this direction, including key organizational changes, noted below. The new leadership of the Bank and the Region is committed to do more to accelerate growth of agriculture and strengthen the sector's contribution to reduction of poverty.
- Drawing on the analysis in the draft *World Development Report 2008: Agriculture for Development* (World Bank 2007c),[1] management notes many areas of agreement. Management also highlights some differences with the IEG review on some of its findings and its recommendations.
- Management notes its view that the IEG review could have usefully given greater weight to issues of country ownership, donor partnerships, alignment with country priorities, and harmonization among donors—key elements of the Paris Declaration and our current and future assistance programs.

The following section elaborates on these points and responds to IEG's recommendations.

Management Views on IEG's Analysis and Conclusions

Management agrees that if Africa is to achieve the MDGs, agriculture must be used more effectively for development. Agriculture still contributes approximately one-third of the aggregate growth of Africa (excluding South Africa). More than two-thirds of Africa's poor people remain in rural areas and depend largely on farming for their livelihoods. As the IEG review notes, African leaders recognize that more must be done to stimulate agricultural growth. They have expressed that view through the Comprehensive African Agricultural Development Program articulated by the New Partnership for Africa's Development (NEPAD). The Bank and other partners are assisting the NEPAD process and helping countries to develop comprehensive programs at the national and regional levels. The Africa Region has identified agricultural productivity as one of eight priority areas. The authors of the *World Development Report* [WDR] *2008: Agriculture for Development* also argue that agriculture can and must contribute more to Africa's development, and they present an agenda for achieving that objective. In that context, management has comments on how, relative to IEG, it sees the wider challenges, and how the Bank has moved to confront these challenges. Management also has comments on several technical areas in the IEG review.

Challenges and recent performance

Management agrees that more must be done. However, strategies to accelerate growth must be built on an understanding of changes already in process. The review presents country-specific data to argue that agriculture's performance has been poor in general.[2] Drawing comprehensively from the same World Development Indicators, the WDR team shows that sectoral growth increased from 2.3 percent in the 1980s to 3.3 percent on average in the 1990s, to 3.8 percent annually in the 2000s—a rate exceeded only by the Middle East and North Africa among the Bank's Regions. The growth derived from a combination of shifts in incentives due to policy changes of the 1990s, higher world commodity prices, area expansion, and changes in the composition of output. Too little of the growth derived from increases in the productivity of land and labor. The WDR team shows that rural poverty rates have started to decline in 10 of the 13 countries for which data were analyzed, but that per capita agricultural growth, although positive on average since 2000, still lags behind that of other Regions and is too slow to meet the MDGs.

Constraints to growth. The IEG review attributes lagging growth to constraints associated with agro-ecological diversity, poor soils, variable rainfall, and frequent droughts. Management agrees but also notes that both technical and broader economic and institutional factors affect performance. Africa's agro-ecology and geography exhibit both positive and negative characteristics; that these aspects have functioned as constraints reflects institutional shortcomings, chronic underinvestment, and residual discrimination in policies. The WDR discusses in some detail the contributions that changes in macroeconomic and sectoral policy made to faster growth over the period and draws attention to the lag structure of cause and effect. Although the policy environment has improved in the past decade and a half, more must be done on the policy and institutional side to facilitate faster growth and improved productivity in African agriculture. Because of the breadth of the agricultural agenda and synergies between and among interventions, the design task presents formidable challenges that go beyond those presented in the review, and that will require a significant analytic investment at both the national and thematic levels. The Bank is working with partners and clients to address these challenges.

The role of aid and the Bank's strategy. The authors of the review note that because of resource constraints, the evaluation considered only the Bank's direct investment lending and nonlending activities. In addition, the evaluation's treatment of the review period as one undifferentiated time slice impedes recognition of change within that period—for example, changes in modalities of assistance. New directions have been taken in the last five years, and the pace of change is accelerating with two significant developments. First, in 2003 African heads of state committed to increase spending on agriculture (from levels rarely in excess of 4 percent of public budgets to 10 percent). Second, the Comprehensive Africa Agriculture Development Program (CAADP) Roundtable process has been established to assist in articulating more robust programs. The nationally owned and regionally supported roundtable process is under way in Benin, Burkina Faso, Burundi, Ethiopia, Ghana, Kenya, Malawi, Mali, Niger, Nigeria, Rwanda, Senegal, Uganda, and Zambia. The programs defined through the roundtables will be financed jointly by national governments, bilateral donors, international financial institutions (including the World Bank), and private foundations. As the programs are articulated and as clients request, the Bank will reflect them in Country Assistance Strategies. With other partners, the Bank is fully engaged in this process both technically and financially, and it is committed to increased engagement. The IEG review's analysis and recommendations could be taken to imply a role for the Bank (notably, increased investment lending through free-standing projects addressing specific technical constraints) that does not sufficiently recognize country leadership, donor alignment, and harmonization.

Organizational changes. To better support a comprehensive and harmonized agenda, the Bank has recently undertaken organizational changes. With the creation of the Sustainable Development Department in fiscal 2007 within the Africa Region, the approximately 80 staff members working on agriculture and rural development have been brought together into one unit under one manager with a mandate to facilitate thematic deployment and shared experience across the Region. Field-based sector leaders within the department have been given the mandate to facilitate coordination across the relevant technical units (agriculture, water, urban, transport, energy, environment, post-conflict, and social). Staff members in the agriculture and rural development unit are engaged in a strategic exercise to clarify best practice in program design and normative costing for five broad thematic areas that constitute building blocks for comprehensive agricultural programs. This work will be completed in fiscal 2008, and will inform the CAADP Roundtables and the Bank's Country Assistance Strategies.

Thematic performance

In a number of technical areas, the authors of the IEG review provide conclusions and recommendations that are not fully consistent with management's views. Key areas of divergence are noted below.

Decentralization. The authors find that decentralization offers little increased scope to address agro-ecological diversity, and that the Bank has done little to support increased productivity in a decentralized environment. Management believes that decentralization holds significant promise, since local governments play an increasing role, and many extension agents now report to local officials rather than central ministries. The IEG review finds that little support has been accorded to efforts to improve productivity, and that food emergencies have instead dominated the Bank's programs. Management is of the view that smallholders are adapting to changing conditions, including planting different varieties and changing planting dates, and that more assistance in this area will bring improved results.

Fertilizer, seeds, and water. The IEG review argues that lack of fertilizer and improved seeds impedes productivity. That is, of course, true, and more must be done. Management points to innovations—some supported under Bank programs—that are being used to address these problems, and also notes important constraints in the policy environment and outside the agricultural sector (such as transport costs) that reduce the profitability of inputs. On irrigation, the authors of the IEG review cite work concluding that irrigated areas can be expanded at acceptable costs. Management draws on the extensive interagency review of past lessons of irrigation in Africa and on analysis presented in the WDR to conclude that increased investments in irrigation are very much needed, but must be identified with careful consideration of the economic and ecological context.

Agricultural finance and extension. The authors of the IEG review see microfinance as a promising remedy for financial constraints that smallholders face. Management acknowledges the importance of microfinance for some producers, but is of the view that it cannot provide the mainstay of rural finance, particularly for agricultural production. Management sees promise in other mechanisms, such as interlocking contracts in the value chain, mobilization of local savings, and such innovative products as e-credit cards for purchasing inputs. With regard to agricultural extension, the IEG review is critical of the training-and-visit approach, and argues that no viable alternative has yet emerged. Management is less pessimistic, and points out that new, demand-responsive approaches have been sufficiently tested on the ground—for example—and offer promise.

Land rights. Management agrees that land rights are important. The authors of the IEG review observe that time taken to complete tenure reforms is often underestimated. Management is of the view that formal recognition of customary tenure, simple documentation of rights, stronger mechanisms for resolution of disputes, and an emphasis on women's land rights can all contribute to increased productivity.

Main Findings and Recommendations

Management finds that IEG's recommendations address issues that are important for advancing agriculture in Sub-Saharan Africa but are not readily translatable into operational actions. Management's specific responses to IEG's recommendations are provided in the attached Management Action Record.[3] Management's commitment to make support for African agriculture a priority is already on record, and has recently been emphasized in assurances from the new vice president of the Africa Region.

The upcoming IEG review. Management notes that since this evaluation serves as a pilot for the proposed IEG review of Bank-wide assistance in agriculture scheduled for fiscal 2009, IEG may wish to consider several points in reference to that future work: (a) the evaluation should be based on the strongest possible analysis; (b) the task should be staffed and funded commensurately with its technical challenge; and (c) the harmonized framework in which the Bank works should be reflected in the design of the assessment.

Management Action Record

<table>
<tr><th>Major IEG recommendation</th><th>Management response</th></tr>
<tr><td>To effectively support the implementation of the Africa Action Plan and its appropriate focus on agricultural development as a key priority, IEG recommends that the Bank:

Focus attention to achieve improvements in agricultural productivity:
• Establish realistic goals for expansion of irrigation and recognize the need to increase productivity of rain-fed agriculture through improvements in land quality, as well as water and drought management.
• Help design efficient mechanisms, including public-private partnerships, to provide farmers with critical inputs including fertilizers, water, credit, and seeds.
• Support the development of marketing and transport infrastructure.</td><td>Agreed with the following qualifications:
Management agrees to take the following actions, which fit with the spirit of IEG's recommendations but are grounded in the world of country ownership, alignment, and harmonization in which the Bank works:
• The Region has developed a draft Business Plan for Irrigation that identifies priority countries and projects in which irrigation can be expanded in the period fiscal 2008–12. The plan will be discussed with partners and, with their agreement, embedded in future Country Assistance Strategies.
• Through agricultural technology operations and support to farmer organizations, the Region is supporting matching grants and various forms of smart subsidies for purchase of improved inputs from the private sector. It is using policy-based operations to assist countries with the regulatory framework for input supply.
• In fiscal 2007, executive directors approved more than $2.5 billion in new lending to Sub-Saharan Africa for infrastructure, much of which serves rural areas. Management notes that, for example, it is now common practice for African governments seeking support from the Bank in the transport sector to involve the agriculture ministry in identifying priority investments. Going forward, explicit attention will continue to be given in fiscal 2008–10 to synergies between infrastructure and commercial agriculture in Bank support.</td></tr>
<tr><td>Improve its work on agriculture:
• Increase the quantity and quality of analytical work on agriculture and ensure that policy advice and lending are grounded in its findings.
• Support public expenditure analyses to assess resource availability for agriculture and to help set Bank priorities.
• Rebuild its technical skills, based on a comprehensive assessment of current gaps.</td><td>Agreed with the following qualifications:
The priority in Africa is assistance to countries in developing their own comprehensive programs to improve agricultural productivity.
• The Bank will concentrate on assisting governments to design, cost, and mobilize the resources needed to implement comprehensive programs of agricultural development.
• As an analytic priority, by fiscal 2010 the Region will assist clients in at least five countries to assess agricultural public expenditure and identify expenditure gaps relative to growth targets for the sector. This work is innovative and methodologically complex, and will be pursued jointly with the Agriculture and Rural Development Department and with other development partners and in the context of the CAADP Roundtable process.
• The Region regularly reviews skill needs and gaps. At present the major gap is in knowledge of issues related to sectoral public spending, both operational practices and analysis.</td></tr>
</table>

(Continues on the following page.)

Management Action Record (*continued*)

Major IEG recommendation	Management response
Establish benchmarks for measuring progress: • Improve data systems to better track activities supported by the Bank. • Strengthen M&E to report on project activities in various agro-ecological zones and for different crops and farmer categories, including women. • Develop a system to coordinate agricultural activities in a country with road access, market proximity, and soil conditions.	***Agreed with the following qualifications:*** • All projects from the Africa Region presented to the Board for approval now require baseline data for the results framework. • A set of core indicators at the national level has been developed to monitor progress under the Africa Action Plan (AAP). For agriculture, these indicators include five-year moving average data on growth in agricultural GDP, productivity per hectare, and productivity per worker. • The Region is experimenting with M&E systems based on GIS (Geographic Information Systems) tracking of data. This may allow collection of information on agro-ecological zones and links with transport. All of the actions noted above in response to the three IEG recommendations are elements of the agricultural productivity component of the AAP (one of the flagship operational constituents of the AAP). Management will consider its agreed actions complete with their successful implementation over the next three years. Management will report agricultural productivity results to executive directors as part of overall monitoring and reporting on the AAP.

Réponse de la direction

La direction approuve et applique déjà le principal message de cet examen — à savoir que la Banque mondiale devrait augmenter son appui à l'agriculture en Afrique subsaharienne. La direction a des observations à faire sur plusieurs éléments de l'analyse de l'IEG et ne partage pas certains de ses avis sur la manière d'atteindre les objectifs communs. Les principaux points sont les suivants :

- La direction est d'accord sur le fait que, pour atteindre les objectifs de développement pour le Millénaire (ODM) en Afrique subsaharienne, l'agriculture doit jouer un rôle plus déterminant. La Région, en collaboration étroite avec les dirigeants africains et les partenaires du développement, a déjà réorienté notablement le travail de la Banque dans cette direction, en procédant notamment à une importante réorganisation, évoquée plus loin. Les nouveaux responsables de la Banque et de la Région sont résolus à accroître leurs efforts pour accélérer la croissance de l'agriculture et renforcer la contribution de ce secteur à la réduction de la pauvreté.
- S'appuyant sur l'analyse du projet de « Rapport sur le développement dans le monde 2008 : l'Agriculture au service du développement » (Banque mondiale 2007c)[1], la direction note de nombreux domaines où les avis se rejoignent. Elle attire également l'attention sur quelques différences avec le rapport de l'IEG au sujet de certaines de ses conclusions et de ses recommandations.
- La direction estime que l'examen de l'IEG aurait pu utilement faire une plus large place aux questions de l'adhésion nationale, des partenariats entre donateurs, de l'alignement de l'aide sur les priorités nationales, et de l'harmonisation de l'aide des donateurs, qui constituent des aspects essentiels de la Déclaration de Paris et de nos programmes d'aide actuels et futurs.

La section suivante approfondit ces points et répond aux recommandations de l'IEG.

Les vues de la direction sur l'analyse et les conclusions de l'IEG

La direction adhère à l'idée que l'agriculture doit être davantage mise au service du développement si l'on veut que l'Afrique atteigne les ODM. Ce secteur contribue encore pour environ un tiers à la croissance globale en Afrique (hors Afrique du Sud). Plus des deux tiers de la population pauvre d'Afrique habite encore en milieu rural et est largement tributaire de l'agriculture pour sa subsistance. Comme le note le rapport de l'IEG, les dirigeants africains reconnaissent qu'il faut accroître les efforts pour stimuler la croissance agricole. Ils ont exprimé cet avis à travers le Programme intégré pour le développement de l'agriculture en Afrique, élaboré dans le cadre du Nouveau partenariat pour le développement de l'Afrique (NEPAD). La Banque et les autres partenaires accompagnent le processus du NEPAD et aident les pays à mettre au point des programmes intégrés à l'échelon national et régional. La productivité agricole fait partie des huit domaines prioritaires identifiés par la Région Afrique. Les auteurs du « Rapport sur le développement dans le monde 2008 : l'agriculture au service du développement » font également valoir que l'agriculture peut et doit contribuer davantage au développement de l'Afrique, et ils présentent un programme d'action pour atteindre cet objectif.

Dans ce contexte, la direction a des observations à formuler sur sa vision de la problématique générale, par rapport à l'IEG, et sur la manière dont la Banque a réorienté son travail pour y faire face. Elle a aussi des commentaires à faire sur plusieurs domaines techniques de l'examen de l'IEG.

Difficultés et résultats récents

La direction convient qu'il faut faire davantage. Toutefois, élaborer des stratégies pour accélérer la croissance suppose de comprendre les changements déjà en cours. Le rapport de l'IEG présente des données par pays pour démontrer que les performances de l'agriculture sont faibles d'une manière générale[2]. En se fondant sur les mêmes Indicateurs du développement dans le monde, l'équipe du RDM montre que le taux de croissance du secteur a progressé, passant de 2,3 % dans les années 80, à 3,3 % en moyenne dans les années 90, puis à 3,8 % par an après l'an 2000 — niveau dépassé seulement par le Moyen-Orient et l'Afrique du Nord parmi les Régions de la Banque. Cette croissance est imputable à une série d'évolutions des dispositifs d'incitation provoquées par des changements dans la politique publique au cours des années 90, à la hausse des cours des produits de base, à l'augmentation des surfaces cultivées, et à des changements dans la composition de la production. Une part trop faible de la croissance a été générée par des augmentations de la productivité des sols et du travail. L'équipe du RDM indique que les taux de pauvreté rurale ont commencé à décliner dans 10 des 13 pays pour lesquels des données ont été analysées, mais que la croissance agricole par habitant, bien que positive en moyenne depuis 2000, reste loin derrière celle des autres Régions et est trop lente pour permettre d'atteindre les ODM.

Obstacles à la croissance. Dans son examen, l'IEG attribue le retard de croissance à des obstacles liés à la diversité agroécologiques, à la pauvreté des sols, à la variabilité des précipitations et à la fréquence élevée des sécheresses. La direction en convient, mais fait remarquer également que les performances de l'agriculture sont affectées par des paramètres techniques et par des facteurs économiques et institutionnels plus généraux. Les caractéristiques agroécologiques et géographiques de l'Afrique présentent autant d'aspects positifs que négatifs ; qu'ils aient constitué un frein à la croissance est la conséquence de carences institutionnelles, d'un sous-investissement chronique et de politiques comportant encore des éléments discriminatoires. Le RDM examine de façon relativement détaillée en quoi les changements introduits dans la politique macroéconomique et sectorielle ont contribué à accélérer la croissance pendant la période considérée, et attire l'attention sur le décalage temporel qui existe entre une cause et ses effets. Bien que le cadre de l'action publique se soit amélioré depuis une quinzaine d'années, il faut renforcer les efforts dans le domaine de la politique publique et des institutions pour augmenter les chances d'accélérer la croissance et d'intensifier la productivité du secteur agricole en Afrique. L'immense éventail des actions à mener dans le domaine de l'agriculture et les synergies qui existent entre et à l'intérieur des interventions rendent la tâche extrêmement difficile à organiser. Les problèmes rencontrés dépassent ainsi le cadre de l'examen de l'IEG et nécessiteront des travaux d'analyse importants sur les pays et sur des thèmes particuliers. La Banque travaille à résoudre ces problèmes avec ses partenaires et ses clients.

Le rôle de l'aide et la stratégie de la Banque. Les auteurs du rapport notent que pour des raisons de moyens, l'évaluation n'a porté que sur les prêts directs à l'investissement et les activités hors prêt. En outre, le fait que la période examinée soit traitée en un seul bloc ne permet pas de tenir compte des changements intervenus au cours des années considérées — par exemple dans les modalités de fourniture de l'aide. De nouvelles orientations ont été prises depuis cinq ans, et le rythme du changement s'accélère avec deux faits marquants. Tout d'abord, en 2003, les chefs d'État africains se sont engagés à augmenter les montants alloués à l'agriculture (qui passeront à 10 % des budgets publics alors qu'ils dépassaient rarement les 4 %). Ensuite, le processus de consultation du Programme intégré pour le développement de l'agriculture en Afrique

(CAADP) a été mis en place pour aider à la formulation de programmes plus solides. Le processus de consultation, pris en charge par les pays avec un soutien régional, est engagé au Bénin, au Burkina Faso, au Burundi, en Éthiopie, au Ghana, au Kenya, au Malawi, au Mali, au Niger, au Nigéria, en Ouganda, au Rwanda, au Sénégal, et en Zambie. Les programmes définis lors de ces tables rondes seront financés conjointement par les gouvernements nationaux, des donateurs bilatéraux, des institutions financières internationales (notamment la Banque mondiale) et des fondations privées. En fonction de la formulation des programmes et de la demande des clients, la Banque en tiendra compte dans ses stratégies d'aide-pays. Avec d'autres partenaires, la Banque s'est totalement impliquée dans ce processus, tant sur le plan technique que financier, et est déterminée à accroître ses engagements. L'analyse et les recommandations formulées par l'IEG dans son rapport pourraient laisser entendre que la Banque aurait un rôle à jouer (notamment, en augmentant les prêts à l'investissement dans le cadre de projets autonomes ciblés sur des problèmes techniques particuliers) qui ne laisserait pas une place suffisante à l'adhésion des pays et à l'alignement et l'harmonisation de l'aide des donateurs.

Réorganisation. Pour mieux appuyer un programme d'action intégré et harmonisé, la Banque a lancé récemment une réorganisation de ses services. Avec la création du Département du développement durable durant l'exercice 07 au sein de la Région Afrique, les quelque 80 collaborateurs qui travaillaient à l'agriculture et au développement rural ont été rassemblés en une seule unité dirigée par une seule personne afin de faciliter l'affectation des compétences et les échanges d'expérience au niveau de la Région. Au sein du département, des responsables de secteur basés sur le terrain ont été chargés d'organiser la coordination entre les équipes techniques concernées (agriculture, eau, urbanisme, transport, énergie, environnement, post-conflit et social). Les membres de l'unité « agriculture et développement rural » ont entamé un exercice stratégique dans l'optique d'identifier les pratiques optimales en matière d'élaboration des programmes et de principes d'établissement des coûts pour les cinq grands domaines thématiques sur lesquels est construit tout programme agricole intégré. Ce travail sera achevé durant l'exercice 08 et servira de guide lors des tables rondes du CAADP et dans les stratégies d'aide-pays de la Banque.

Résultats par domaines

Dans un certain nombre de domaines techniques, les auteurs du rapport de l'IEG présentent des conclusions et des recommandations qui ne sont pas totalement en phase avec les opinions de la direction. Les principaux points de divergence sont exposés ci-dessous :

Décentralisation. Les auteurs concluent que la décentralisation ne permet pas d'améliorer sensiblement la prise en compte de la diversité agroécologique, et que la Banque n'a pas fait grand-chose pour favoriser l'augmentation de la productivité dans un cadre décentralisé. La direction estime que la décentralisation offre des perspectives très prometteuses dans la mesure où les autorités locales jouent un rôle grandissant et où un grand nombre d'agents de vulgarisation ne rendent maintenant plus compte aux ministères centraux mais aux fonctionnaires locaux. Le rapport de l'IEG pointe le peu d'appui accordé aux efforts d'amélioration de la productivité et indique que les programmes de la Banque ont privilégié l'aide alimentaire d'urgence. La direction est d'avis que les petits exploitants s'adaptent aujourd'hui au changement, par exemple en plantant des variétés différentes et en modifiant les dates de semis, et qu'un surcroît d'assistance dans ce domaine permettra d'obtenir de meilleurs résultats.

Engrais, semences et eau. Dans son rapport, l'IEG indique que le manque d'engrais et de semences améliorées est un frein à la productivité. C'est bien entendu exact, et il faut être plus actif dans ce domaine. La direction signale les solutions originales qui ont été trouvées pour y remédier — dont certaines sont financées par des programmes de la Banque — et note également les obstacles majeurs qui existent au niveau du cadre de la politique publique et en dehors du secteur agricole (par exemple les coûts de transport), qui diminuent les gains générés par les intrants. Au sujet de l'irrigation, les auteurs du rapport de l'IEG citent des travaux concluant à la possibilité d'étendre les périmètres irrigués pour un coût acceptable. La direction, se fondant sur la vaste analyse inter-agence conduite sur l'expérience passée de l'irrigation en Afrique, ainsi que sur l'analyse présentée dans le RDM, conclut qu'il est absolument nécessaire d'accroître les investissements dans l'irrigation, mais de façon ciblée, après un examen approfondi du contexte économique et écologique.

Finance agricole et vulgarisation. Les auteurs du rapport de l'IEG voient dans la microfinance une solution prometteuse aux handicaps financiers des petits exploitants. La direction reconnaît l'importance de la microfinance pour certains producteurs, mais estime qu'elle ne peut pas constituer l'élément essentiel de la finance rurale, en particulier pour la production agricole. La direction met beaucoup d'espoir dans d'autres mécanismes, comme les contrats réciproques dans la chaîne de valeur, la mobilisation de l'épargne locale, et des produits novateurs tels que les porte-monnaie électroniques pour l'achat des intrants. En ce qui concerne la vulgarisation, le rapport de l'IEG est critique vis-à-vis de la méthode « formation et visites », et estime qu'aucune autre solution viable ne s'est pour l'instant dégagée. La direction est moins pessimiste et indique que de nouvelles approches, fondées sur la demande, ont été suffisamment testées sur le terrain — par exemple — et sont prometteuses.

Droits fonciers. La direction est d'accord sur le fait que les droits fonciers sont importants. Les auteurs du rapport de l'IEG relèvent que l'on sous-estime souvent le temps nécessaire pour mener à bien les réformes foncières. La direction considère qu'un certain nombre de mesures peuvent toutes contribuer à accroître la productivité, notamment la reconnaissance officielle de l'occupation coutumière, la simplification des procédures d'immatriculation sur le plan des documents à fournir, le renforcement des mécanismes de règlement des différends et une plus grande attention donnée aux droits fonciers des femmes.

Principales conclusions et recommandations

La direction estime que les recommandations de l'IEG portent sur des questions importantes pour faire avancer l'agriculture en Afrique subsaharienne mais qui ne sont pas directement traduisibles en mesures opérationnelles. Les réponses précises de la direction aux recommandations de l'IEG sont présentées dans le Relevé des interventions de la direction ci-joint[3]. La volonté de la direction de faire de l'aide à l'agriculture africaine l'une de ses priorités est déjà établie et a été soulignée récemment dans les déclarations du nouveau Vice-président de la Région Afrique.

Prochain examen de l'IEG. La direction indique que, puisque cette évaluation doit servir de pilote à l'analyse de l'aide de la Banque mondiale à l'agriculture en général que l'IEG prévoit d'effectuer durant l'exercice 09, l'IEG pourrait utilement examiner plusieurs points dans la perspective de ce travail futur : a) l'évaluation devrait être fondée sur l'analyse la plus approfondie possible ; b) les moyens humains et financiers mis en œuvre devraient être à la mesure du défi technique que représente la tâche à accomplir ; et c) la structure de l'évaluation devrait tenir compte du cadre harmonisé dans lequel travaille la Banque.

Relevé des interventions de la direction

Recommandation de l'IEG	Réponse de la direction
Pour appuyer utilement la mise en œuvre du Plan d'action pour l'Afrique et l'objectif prioritaire de développement agricole qu'il s'est judicieusement fixé, l'IEG formule à l'intention de la Banque mondiale les recommandations suivantes : **Privilégier l'amélioration de la productivité agricole :** • Établir des objectifs réalistes en matière d'expansion de l'irrigation et prendre conscience de la nécessité d'augmenter la productivité de l'agriculture non irriguée par l'amélioration de la qualité des terres et par la gestion de l'eau et des sécheresses. • Participer à l'élaboration de mécanismes efficients, dont des partenariats public-privé, pour fournir aux agriculteurs les intrants indispensables, notamment des engrais, de l'eau, des crédits et des semences. • Appuyer le développement des infrastructures de commercialisation et de transport.	***Acceptée avec les précisions suivantes :*** La direction accepte de prendre les mesures suivantes, qui sont conformes à l'esprit des recommandations de l'IEG mais sont fondées sur les principes de l'adhésion nationale et de l'alignement et de l'harmonisation de l'aide sur lesquels repose le travail de la Banque. • La Région a élaboré un projet de Plan stratégique pour l'irrigation qui identifie les pays et les projets prioritaires dans lesquels l'irrigation peut être étendue pendant la période comprise entre les exercices 08 et 12. Le plan sera examiné avec les partenaires et, avec leur accord, intégré dans les futures stratégies d'aide-pays. • Dans le cadre de ses activités d'assistance technique à l'agriculture et d'aide aux associations d'agriculteurs, la Région finance des subventions de contrepartie et diverses formes de subventions intelligentes à l'achat d'intrants de meilleure qualité auprès du secteur privé. Elle utilise les opérations à l'appui de réformes pour aider les pays à organiser un cadre réglementaire adapté pour les achats d'intrants. • Durant l'exercice 07, les Administrateurs ont approuvé plus de 2,5 milliards de nouveaux prêts à l'Afrique subsaharienne pour des infrastructures dont une bonne part desservent des zones rurales. La direction note que, par exemple, les gouvernements africains font maintenant fréquemment appel à la Banque dans le secteur des transports pour engager le ministère de l'Agriculture à identifier les investissements prioritaires. À l'avenir, une attention particulière continuera d'être accordée aux synergies entre les infrastructures et l'agriculture commerciale dans l'aide apportée par la Banque sur les exercices 08 à 10.
Améliorer ses études dans le domaine de l'agriculture : • Augmenter la quantité et la qualité des études analytiques portant sur l'agriculture et veiller à ce que ses opérations de conseil et de prêt se fondent sur leurs conclusions. • Appuyer les analyses des dépenses publiques de manière à évaluer les ressources disponibles pour l'agriculture et à définir ses propres priorités. • Restaurer ses compétences techniques à partir d'une évaluation complète de ses lacunes actuelles.	***Acceptée avec les précisions suivantes :*** La priorité en Afrique est d'aider les pays à élaborer leurs propres programmes intégrés pour améliorer la productivité agricole. • La Banque s'emploiera avant tout à aider les autorités nationales à élaborer, chiffrer et mobiliser les ressources nécessaires à la mise en œuvre de programmes intégrés de développement agricole. • Dans le domaine des études, d'ici l'exercice 10, la priorité de la Région sera d'aider les clients d'au moins cinq pays à évaluer les dépenses publiques dans l'agriculture et à repérer les lignes budgétaires insuffisantes par rapport aux ob-

Relevé des interventions de la direction (*suite*)

Recommandation de l'IEG	Réponse de la direction
	jectifs de croissance définis pour le secteur. Il s'agit d'un travail nouveau et complexe sur le plan méthodologique, qui sera mené en collaboration avec le Département de l'agriculture et du développement rural et d'autres partenaires du développement, et dans le contexte du processus de consultation du CAADP. • La Région fait régulièrement le point sur les besoins et les déficits de compétences. Actuellement, les principales lacunes concernent la connaissance des questions liées aux dépenses publiques sectorielles, tant du point de vue des pratiques opérationnelles que de l'analyse.
Établir des indicateurs afin de mesurer les progrès : • Perfectionner les systèmes de données pour mieux suivre les opérations que la Banque finance. • Consolider le dispositif de suivi et d'évaluation pour rendre compte des activités conduites dans le cadre des projets dans diverses zones agroécologiques, et pour différentes cultures et catégories d'agriculteurs, y compris les femmes. • Mettre sur pied un système permettant de coordonner les opérations agricoles à l'échelle des pays en fonction de leur réseau routier, de la proximité des marchés et des conditions édaphiques.	***Acceptée avec les précisions suivantes :*** • Tous les projets de la Région Afrique présentés au Conseil pour approbation doivent maintenant comporter des données de référence pour le cadre de résultats. • Un ensemble d'indicateurs clés au niveau national a été élaboré pour suivre les progrès accomplis dans le cadre du Plan d'action pour l'Afrique (PAA). Pour l'agriculture, ces indicateurs comprennent des données en moyenne glissante sur cinq ans sur la croissance du PIB agricole, la productivité à l'hectare et la productivité par employé. • La Région expérimente des systèmes de suivi et d'évaluation fondés sur le suivi de données de SIG (système d'information géographique), qui pourraient permettre de collecter des informations sur les zones agroécologiques et d'établir des liens avec le secteur des transports. Toutes les interventions indiquées ci-dessus en réponse aux trois recommandations de l'IEG s'inscrivent dans la composante de productivité agricole du PAA (l'une des mesures opérationnelles phares du PAA). La direction considérera que les actions qu'elle a convenu de poursuivre ont été menées à leur terme si elles sont mises en œuvre avec succès au cours des trois prochaines années. La direction informera les Administrateurs des résultats obtenus en matière de productivité agricole dans le cadre du dispositif global de suivi et de communication de données du PAA.

A Resposta da Administração

A Administração concorda e já está a por em prática a principal mensagem da presente análise, isto é, que o Banco Mundial deveria prestar mais apoio à agricultura na África Subsariana. A Administração tem alguns comentários a fazer sobre vários elementos da análise do Grupo Independente de Avaliação (GIA), e difere de algumas das opiniões do GIA quanto à maneira de proceder para alcançar os objectivos comuns. Os principais pontos são:

- A Administração concorda que, para alcançar os Objectivos do Milénio para o Desenvolvimento (OMD) na África Subsariana, a agricultura tem que desempenhar um papel mais efectivo. A Região, em colaboração estreita com os dirigentes africanos e os parceiros que se dedicam ao desenvolvimento, já dirigiu o Banco de uma maneira significativa nessa direcção, o que inclui as importantes mudanças organizativas abaixo indicadas. Os novos dirigentes do Banco e a Região comprometeram-se a fazer mais no sentido de acelerar o crescimento da agricultura e de reforçar a contribuição do sector para reduzir a pobreza.
- Tendo por base a análise contida no *Relatório sobre o Desenvolvimento Mundial de 2008: A Agricultura para o Desenvolvimento* (Banco Mundial 2007c),[1] a Administração verifica que há acordo em muitas áreas. A Administração salienta também algumas divergências em relação a algumas das conclusões e recomendações da análise do GIA.
- A Administração expressa a opinião de que a análise do GIA poderia ter dado mais peso às questões relacionadas com o envolvimento dos países, às parcerias com os doadores, ao alinhamento às prioridades dos países, e à harmonização entre os doadores, o que teria sido útil visto estes serem elementos fundamentais da Declaração de Paris e dos nossos programas de assistência actuais e futuros.

A parte seguinte contém uma descrição mais pormenorizada destes pontos e a resposta às recomendações do GIA.

A Opinião da Administração sobre a Análise e as Conclusões do GIA

A Administração concorda com o facto que, para que a África alcance os OMD, a agricultura tem que ser usada mais efectivamente em prol do desenvolvimento. A agricultura continua a contribuir com aproximadamente um terço para o crescimento agregado da África (com a exclusão da África do Sul). Mais de dois terços das pessoas pobres na África continuam a viver nas zonas rurais e dependem em grande medida da agricultura para a sua sobrevivência. Como indicado na análise do GIA, os dirigentes africanos reconhecem que é preciso fazer mais para estimular o crescimento agrícola. Eles expressaram esta opinião através do Programa Abrangente da África para da Agricultura, articulado pela Nova Parceria para o Desenvolvimento da África (NPDA-NEPAD). O Banco e os outros parceiros estão a auxiliar o processo da NPDA e a ajudar os países a elaborarem programas abrangentes a nível nacional e regional. A Região da África identificou a produtividade agrícola como sendo uma de oito áreas prioritárias. Os autores do *Relatório Sobre o Desenvolvimento Mundial* [RDM] de *2008: A Agricultura para o Desenvolvimento* também apresentam o argumento que a agricultura pode e deve contribuir mais para o desenvolvimento da África, e propõem uma agenda para alcançar este objectivo. Neste contexto, a Administração tem comentários a fazer em relação ao GIA, à maneira como visualiza os desafios mais amplos, e à maneira como o Banco procedeu para enfrentar estes desafios. A Administração também tem comentários a fazer sobre vários aspectos técnicos da análise do GIA.

Os Desafios e o Desempenho Recente

A Administração concorda que é necessário fazer mais. Contudo, as estratégias para acelerar o crescimento devem assentar na compreensão das mudanças já estão em curso. A análise apresenta dados específicos aos países para argumentar que o desempenho da agricultura tem sido, de uma maneira geral, deficiente.[2] Baseando-se amplamente nos mesmos Indicadores do Desenvolvimento Mundial, a equipa encarregada do RDM mostra que o crescimento sectorial aumentou de 2,3 por cento na década de 1980 para uma média de 3,3 por cento na década de 1990, e para 3,8 por cento anuais na década de 2000 — uma taxa que só foi ultrapassada pelo Médio Oriente e a África do Norte entre todas as Regiões do Banco. O crescimento foi devido a uma combinação de mudanças nos incentivos derivadas das alterações de política da década de 1990, à subida dos preços mundiais dos produtos primários, à expansão das áreas (de cultivo), e às alterações na composição da produção. Muito pouco do crescimento derivou de um aumento da produtividade das terras e da mão-de-obra. A equipa encarregada do RDM mostra que as taxas de pobreza nas zonas rurais começaram a diminuir em 10 dos 13 países cujos dados foram analisados, mas que o crescimento agrícola per capita, embora sendo em média positivo desde o ano 2000, ainda é menor do que o das outras Regiões e é demasiado lento para que os OMD sejam alcançados.

Os Constrangimentos ao Crescimento. A análise do GIA atribui o atraso no crescimento a constrangimentos associados à diversidade agro-ecológica, à pobreza dos solos, à pluviosidade variável, e às secas frequentes. A Administração concorda, mas também assinala que factores tanto técnicos como económicos e institucionais mais amplos afectam o desempenho. A agro-ecologia e a geografia da África exibem características tanto positivas como negativas; o facto que estes factores funcionaram como constrangimentos reflecte as limitações institucionais, o sub-investimento crónico, e a discriminação das medidas de políticas adoptadas. O RDM expõe detalhadamente as contribuições que as mudanças da política macroeconómica e sectorial fizeram para que o crescimento fosse mais rápido durante o período em questão, e dirige a atenção para a estrutura do atraso, de causa e efeito. Embora a conjuntura de política tenha melhorado na última década e meia, é preciso fazer mais na vertente política e institucional para facilitar um crescimento mais rápido e melhorar a produtividade da agricultura da África. Devido à vastidão da agenda da agricultura e às sinergias existentes entre as intervenções, o trabalho de concepção apresenta desafios enormes que vão para além dos que foram apresentados na análise, e que requerem um importante investimento analítico em termos de análise a nível nacional e temático. O Banco está a trabalhar com os seus parceiros e clientes para abordar estes desafios.

A Função das Ajudas e a Estratégia do Banco. Os autores da análise assinalam que, devido aos constrangimentos em matéria de recursos, a avaliação apenas examinou os empréstimos para investimento directo e as actividades não relacionadas com os empréstimos do Banco. Adicionalmente, o tratamento dado na avaliação ao período em análise, como sendo uma fatia de tempo indiferenciada, impede o reconhecimento das mudanças ocorridas durante esse período — por exemplo, a alteração das modalidades de assistência. Foram seguidas novas direcções nos últimos cinco anos, e o ritmo das mudanças está a acelerar com duas ocorrências significativas. Primeiro, em 2003 os chefes de estado africanos comprometeram-se a aumentar os gastos com a agricultura (de níveis que raramente ultrapassaram os 4 por cento do orçamento do estado, para 10 por cento).

Segundo, foi criado o processo de Mesas Redondas do Programa Abrangente para o Desenvolvimento da Agricultura na África (PADAA - CAADP) para ajudar a articular programas mais robustos. O processo de mesas redondas do qual os países assumiram a propriedade e que recebe apoio regional, já está a funcionar no Benin, no Burkina Faso, no Burundi, na Etiópia, no Gana, no Quénia, no Malawi, no Mali, no Niger, na Nigéria, no Ruanda, no Senegal, no Uganda, e na Zâmbia. Os programas definidos através das mesas redondas serão financiados conjuntamente pelos governos dos países, por doadores bilaterais, por instituições financeiras internacionais (incluindo o Banco Mundial), e por fundações privadas. À medida que os programas forem articulados os programas e conforme os clientes solicitarem, o Banco irá reflecti-los nas Estratégias para a Assistência aos Países. Com os outros parceiros, o Banco participa plenamente neste processo, tanto do ponto de vista técnico como financeiro, e compromete-se a ter uma maior participação Poderia considerar-se que a análise e as recomendações do GIA atribuem uma função ao Banco (nomeadamente, um aumento dos empréstimos para investimentos através de projectos independentes que abordem constrangimentos técnicos específicos) que não reconhece suficientemente os dirigentes dos países, o alinhamento dos doadores e a harmonização.

Alterações Organizativas. Para melhor apoiar uma agenda abrangente e harmonizada, o Banco empreendeu recentemente algumas mudanças a nível da organização. Com a criação do Departamento de Desenvolvimento Sustentável no exercício financeiro de 2007 na Região da África, os aproximadamente 80 funcionários que trabalham na agricultura e no desenvolvimento rural foram reunidos numa única unidade, cujo director tem o mandato de facilitar o desdobramento temático e as experiências compartilhadas através da Região. Os chefes do sector dentro do departamento que trabalham no terreno receberam o mandato de facilitar a coordenação entre as unidades técnicas pertinentes (agricultura, água, zonas urbanas, transportes, energia, ambiente, situações posteriores a conflitos, e área social). Os funcionários da unidade de agricultura e desenvolvimento estão a realizar um exercício estratégico para esclarecer as melhores práticas na concepção dos programas e as normas para a determinação dos custos em cinco amplas áreas temáticas que constituem elementos da construção de os programas abrangentes para a agricultura. Este trabalho estará terminado no ano fiscal de 2008 e vai informar as Mesas Redondas do PADAA – CAAD e as Estratégias para a Assistência aos Países do Banco Mundial.

Desempenho Temático

Em algumas áreas técnicas, os autores da análise do GIA apresentam conclusões e recomendações que não são totalmente compatíveis com as opiniões da Administração. Os principais pontos de divergência estão indicados abaixo.

Descentralização. Os autores acham que a descentralização poucas possibilidades oferece de abordar a diversidade agro-ecológica, e que o Banco pouco faz para apoiar uma maior produtividade numa conjuntura de descentralização. A Administração acredita que a descentralização é bastante promissora visto que os governos locais desempenham um papel cada vez maior, e muitos agentes dos serviços de extensão se reportam actualmente às autoridades locais e não aos ministérios centrais. A análise do GIA verifica que pouco apoio foi prestado aos esforços tendentes a melhorar a produtividade, e que em vez disso as situações de emergência alimentar dominaram os programas do Banco. A Administração é de opinião que os pequenos agricultores estão a adaptar-se às condições cambiantes, o que inclui a plantação de variedades diferentes e a mudança das datas de plantação, e que uma assistência acrescida nesta área dará resultados melhores.

Fertilizantes, Sementes e Água. A análise do GIA argumenta que a falta de fertilizantes e de sementes melhoradas impede a produtividade. Isto é evidentemente verdade e é necessário fazer mais a esse respeito. A Administração aponta para as inovações — algumas com o apoio prestado no âmbito dos programas do Banco — que estão a ser utilizadas para abordar esses problemas, e também assinala os constrangimentos importantes na conjuntura de política e fora do sector da agricultura (como os custos dos transportes) que reduzem a lucratividade dos insumos. Quanto à irrigação, os autores da análise do GIA citam o trabalho que levou à conclusão que as áreas irrigadas podem ser ampliadas a um custo aceitável. A Administração baseia-se na longa análise interinstitucional dos ensinamentos passados relativos à irrigação na África e na análise apresentada no RDM que concluía que são absolutamente necessários maiores investimentos na irrigação, mas que estes devem ser identificados levando cuidadosamente em consideração o contexto económico e ecológico.

O Financiamento da Agricultura e a Extensão. Os autores da análise do GIA vêem o microfinanciamento como um remédio promissor para os constrangimentos financeiros que os pequenos agricultores enfrentam. A Administração reconhece a importância do micro financiamento para alguns produtores, mas é de opinião que ele não pode ser o principal financiamento para as zonas rurais, especialmente para a produção agrícola. A Administração considera que outros mecanismos são promissores, tais como ligar os contratos à cadeia de valor, a mobilização da poupança local, e produtos inovadores como os cartões de crédito electrónico para a compra dos insumos. No que respeita a extensão agrícola, a análise do GIA critica a abordagem seguida de formação e visitas, e argumenta que ainda não surgiu nenhuma alternativa viável. A Administração é menos pessimista e indica que as novas abordagens que reagem à procura já foram suficientemente testadas no terreno, por exemplo, e são promissoras.

Direitos de Propriedade das Terras. A Administração concorda que os direitos de propriedade das terras são importantes. Os autores da análise do GIA observam que o tempo que leva a concluir as reformas da propriedade das terras é muitas vezes subestimado. A Administração é de opinião que o reconhecimento oficial da propriedade usual, a documentação simples dos direitos, mecanismos mais fortes para resolver as diferendos, e o ênfase dado ao direito de as mulheres serem proprietárias de terras, todos eles podem contribuir para uma produtividade acrescida.

Principais Conclusões e Recomendações

A Administração acha que as recomendações do GIA abordam assuntos que são importantes para o avanço da agricultura na África Subsariana, mas que não se podem traduzir facilmente em acções operacionais. As respostas específicas da Administração às recomendações do GIA encontram-se em anexo no Registo das Acções da Administração.[3] O comprometimento da Administração de atribuir a prioridade ao apoio à agricultura da África já está patente, tendo sido ressaltado recentemente nas garantias dadas pelo novo Vice-Presidente da Região da África.

A Próxima Análise do GIA. A Administração faz notar que, visto que esta avaliação serve de orientação para a análise proposta do GIA da assistência prestada por todo o Banco à agricultura, programada para o ano fiscal de 2009, talvez o GIA queira ponderar vários aspectos relativos a esse trabalho futuro: (a) a avaliação deveria basear-se na análise mais robusta possível; (b) a dotação de pessoal e o financiamento deste trabalho deveria ser comensurado com aos desafios técnicos enfrentados; e (c) a estrutura harmonizada na qual o Banco trabalha deveria estar reflectida na concepção da avaliação.

Registo das Acções Empreendidas pela Administração

Recomendação do GIA	Resposta da Administração
Para apoiar efectivamente a execução do Plano de Acção para a África e a sua focalização apropriada no desenvolvimento da agricultura como prioridade principal, o GIA recomenda que o Banco: **Concentre a sua atenção na consecução do melhoramento da produtividade agrícola**: • Defina metas realistas para a expansão da irrigação e reconheça a necessidade de aumentar a produtividade da agricultura de sequeiro mediante o melhoramento da qualidade das terras, assim como a gestão dos recursos hídricos e das secas. • Ajude a conceber mecanismos eficientes, incluindo parcerias entre os sectores público e privado, para fornecer aos agricultores os insumos essenciais, inclusivamente fertilizantes, água, crédito e sementes. • Apoie o desenvolvimento das infra-estruturas de comercialização e de transportes.	***Concordou com as qualificações seguintes:*** A Administração concorda em empreender as seguintes acções que se inserem no espírito das recomendações do GIA, mas se baseiam no mundo de propriedade, alinhamento e harmonização dos países nos quais o Banco trabalha. • A Região desenvolveu um Plano Empresarial preliminar para a Irrigação, o qual identifica os países e projectos prioritários nos quais se pode expandir a irrigação no período fiscal de 2008 a 2012. O plano será debatido com os parceiros e, com o assentimento deles, será embutido nas Estratégias para a Assistência aos Países futuras. • Através das operações de tecnologia agrícola e o apoio às organizações de agricultores, a Região está a prestar apoio a doações de contrapartida e várias formas de subsídios "inteligentes" para a compra de insumos melhorados ao sector privado. Está a utilizar as operações baseadas em política para auxiliar os países no que respeita a estrutura regulamentar para a oferta de insumos. • No ano fiscal de 2007, os Administradores do Banco aprovaram mais de USD 2,5 biliões de novos empréstimos à África Subsariana para infra-estruturas, muitas das quais servem as zonas rurais. A Administração assinala, por exemplo, que agora é prática comum para os governos africanos procurar obter apoio do Banco para o sector dos transportes para que o Ministério da Agricultura participe na identificação dos investimentos prioritários. Avançando um pouco, continuará a ser prestada uma atenção explícita nos anos fiscais de 2008 a 2010 às sinergias entre as infra-estruturas e a agricultura comercial no apoio prestado pelo Banco presta.
Realize um trabalho melhor na área da agricultura: • Aumente a quantidade e a qualidade do trabalho analítico relativo à agricultura, e assegure que o assessoramento de política e os empréstimos que concede estão ancorados nas suas conclusões. • Apoie as análises das despesas públicas para avaliar os recursos disponíveis à agricultura, e ajudar a estabelecer as prioridades do Banco. • Reconstrua as suas competências técnicas, com base numa avaliação abrangente das lacunas actuais.	***Concordou com as seguintes qualificações:*** A prioridade na África é a assistência aos países para que eles desenvolvam os seus próprios programas abrangentes destinados a melhorar a produtividade agrícola. • O Banco vai concentrar-se em prestar assistência aos governos destinada a conceber, determinar os custos e mobilizar os recursos necessários para executar programas abrangentes de desenvolvimento da agricultura. • Como prioridade analítica, até ao ano fiscal de 2010 a Região vai auxiliar clientes em pelo menos cinco países a avaliar as despesas públicas com a agricultura e a identificar a brecha nas despesas em relação às metas de crescimento

Registo das Acções Empreendidas pela Administração

Recomendação do GIA	Resposta da Administração
	para o sector, Este trabalho é inovador e complexo, do ponto de vista metodológico, e será realizado conjuntamente com o Departamento da Agricultura e do Desenvolvimento Rural e com outros parceiros no desenvolvimento, e no contexto do processo de Mesas Redondas do PADAA – CAADP. • A Região examina regularmente as necessidades em matéria de competências e as lacunas existentes. No momento actual, a maior lacuna é no conhecimento dos assuntos relacionados com as despesas públicas sectoriais, tanto as práticas operacionais como as análises.
Estabeleça referências para medir os progressos: • Melhore os sistemas de dados para fazer melhor o seguimento das actividades que recebem apoio do Banco. • Reforce os serviços de Monitoramento e Avaliação (M&A) para que informem sobre as actividades dos projectos nas diversas zonas agro-ecológicas e para diferentes categorias de cultivos e agricultores, incluindo as mulheres. • Desenvolva um sistema para coordenar as actividades agrícolas num país com o acesso rodoviário, a proximidade dos mercados e as condições dos solos.	***Concordou com as seguintes qualificações***: • Todos os projectos da Região da África que são apresentados à aprovação do Conselho de Administração têm agora que conter os dados básicos para a estrutura de resultados. • Foi desenvolvido um conjunto de indicadores básicos a nível nacional para monitorar os progressos realizados no âmbito do Plano de Acção para a África (PAA). No que respeita a agricultura, estes indicadores incluem dados médios para cinco anos sobre o crescimento do PIB na agricultura, a produtividade por hectare e a produtividade por trabalhador. • A Região está a experimentar os sistemas de Monitoramento e Avaliação (M&A) baseados nos Sistemas de Informação Geográfica (SIG) que fazem o seguimento dos dados. Isto pode permitir recolher informações sobre as zonas agro-ecológicas e as suas ligações aos transportes. Todas as acções acima indicadas, em resposta às três recomendações do GIA, são elementos da componente de produtividade agrícola do PAA (um dos constituintes operacionais capitais do PAA). A Administração vai considerar as acções que concordou empreender, juntamente com a boa execução das mesmas durante os próximos três anos. A Administração informará os Administradores do Banco sobre os resultados da produtividade agrícola como parte das informações e do monitoramento global sobre o PAA.

Chairperson's Summary: Committee on Development Effectiveness (CODE)

On October 3, 2007, the Committee on Development Effectiveness (CODE) considered the report *World Bank Assistance to Agriculture in Sub-Saharan Africa: An IEG Review*, prepared by the Independent Evaluation Group (IEG), together with the Draft Management Response.

Background

CODE considered the IEG evaluation *World Bank Support for Capacity Building in Africa* on March 23, 2005. IEG also prepared the special study *The World Bank's Assistance to Agricultural Water Management* (1994–2004), dated July 7, 2006.

The draft *World Development Report 2008: Agriculture for Development* was discussed by the Board on June 12, 2007, and is scheduled to be launched on October 19. The main message is that agriculture remains a fundamental instrument for development, and that in Africa this requires addressing constraints to agricultural productivity among smallholders—both those engaged primarily in subsistence and those already commercially active.

IEG Report

The IEG review was primarily meant to provide timely evaluation insight into the implementation of the Bank's renewed focus on agriculture in Africa, especially as embodied in the Africa Action Plan. It will also help design the framework for the proposed IEG study of Bank-wide assistance in agriculture scheduled for fiscal 2009.

The study assesses the development effectiveness of World Bank assistance in addressing constraints to agricultural development in Sub-Saharan Africa over the period of fiscal 1991–2006. The study notes that the agricultural sector has been neglected both by governments and the donor community, including the World Bank. It finds that the Bank's limited and, until recently, declining support has been largely piecemeal and "sprinkled" across several critical areas such as research, extension, credit, seeds, roads, and policy reforms, but with little recognition of the synergy between them. As a result, though there have been areas of comparatively greater success—research, for example—results on the ground have been limited because of weak linkage with extension and limited availability of critical complementary factors such as fertilizers, water, and market access. Poor governance and conflict in several countries, weak institutional capacity, and inadequate government appreciation of the importance of agriculture in development, as well as insufficient coordination of donor efforts, have been factors in the continent's poor agricultural performance.

The study also finds that the Bank's technical skills to support agricultural development have declined over time and that its analytical work has been limited, of variable quality, and has not strategically informed lending program design or policy advice. The study recommends that the Bank should: (i) Support improvements in agricultural productivity by helping design mechanisms that can bring various factors such as improved seeds, water, credit, and good extension advice, among others, to farmers in a coordinated manner; (ii) Focus on improving its own capacity to adequately support agricultural development by increasing the quantity and quality of analytical work to help set country-level priorities and ensure that policy advice and lending are grounded in its findings and by rebuilding its technical skills; (iii) Improve its data systems to better track activities supported by the Bank and strengthen M&E to report accurately on project activities in various countries.

Draft Management Response

Management agrees with the broad directions of IEG's findings and recommendations and is already putting many of them in place. Management's ongoing strategic exercise to identify and dimension comprehensive agricultural programs informs the CAADP Roundtables and assists our clients in designing their own programs. Management differs from IEG in some areas of emphasis. The IEG review attributes lagging growth to constraints associated with agro-ecological diversity, poor soils, variable rainfall, and frequent droughts. Management agrees, but also notes that both technical and broader economic and institutional factors affect performance. Management finds that IEG's recommendations address issues that are important for advancing agriculture in Africa but that several specific recommendations (for example, assure timely access to inputs) require further elaboration before they are actionable. Management is supporting the needed elaboration under the ongoing assistance to countries in the definition of comprehensive programs. Management differs from IEG in placing greater weight on country ownership of programs and harmonization with other development partners.

Overall Conclusions

Members welcomed the timely discussion of this study and on agriculture in Africa, particularly its anticipated inclusion as a main message of the president for the Annual Meetings, and the renewed emphasis on agricultural development in economic growth and poverty reduction. Members found the discussion complemented well the 2008 *World Development Report* (WDR), Africa Action Plan, and work toward the MDGs. Members stressed the need to ensure that the Bank has an integrated agricultural approach to address systemic issues, identifying binding constraints within different country contexts, and addressing urgent needs of African poor countries with a multisectoral and multifaceted approach. There was consensus on the need for realistic goals for Bank involvement and appropriate balance between analytical work and policy advice. Directors also supported the efforts to revitalize the Bank's engagement based on its comparative advantages in promoting agricultural development, but with clear coordination and cooperation with other development partners such as the FAO.

Members commented on the need to include governments' views, define the public sector role vis-à-vis the private sector, and adapt to the global aid architecture and advance the implementation of the Paris Declaration, avoiding fragmentation of assistance. They also stressed the importance of addressing governance issues, corruption, and land tenure, while recognizing the social, economic, and political sensitivity of this issue. In this vein, while agreeing on the need for a country-owned and demand-driven approach, members expressed diverse opinions about the Bank's role and involvement. Members also stressed the adaptation to climate change, access to credit for small farmers, and development of innovative financial instruments, particularly for risk management, as well as the importance of addressing the trade agenda. Other issues that resonated with several participants were: financial resources needed, use of trust funds, and the role of IDA lending and grants; development of technology and technology transfer; categorization of countries, taking into account specific levels of development; institutional capacity and sustainability of policy reforms; and cross-sectors such as infrastructure, transport, water, access to markets, and the gender dimension.

Next Steps

There was a request for an update on the Rural Development Sector Strategy, which management indicated had been last discussed in the 2005 Sector Strategy Implementation Update. Members looked forward to considering the proposed IEG study on Bank-wide assistance in agriculture in fiscal 2009, although a few speakers requested earlier consideration of this report.

The following main issues were raised during the meeting:

The agriculture sector in Africa

Members stressed the importance of the agriculture sector for Africa, its contribution to growth and poverty reduction, and in reaching the Millennium Development Goals (MDGs). They agreed with IEG that agriculture was largely neglected by governments and donors, and noted that the Bank's investment lending to the sector has been relatively low. A few members noted that the agriculture problem goes beyond the Africa Region. One of them sought further information on outstanding lending volume. *Management responded that the Bank's Africa agriculture portfolio of 49 projects is $2.2 billion. New loans, credits, and grants for the last two years (fiscal 2006 and 2007) exceeded $550 million each year, an 80 percent increase compared with the average for fiscal 2001 to 2005.*

Bank contribution

There were comments on the need for enhanced and scaled-up Bank support for agriculture in Africa, a request for a proposal for streamlining the Bank's engagement, and a status report on activities for raising agricultural productivity in the Region. A member emphasized the need to consider agriculture in Africa in the context of the World Bank Group's Long-Term Strategy Exercise, led by the chief economist. Relatedly, the Bank Group should maintain coordination and consistency in country programming. A question was raised on whether there was a need to look at organizational aspects and management systems in the Bank.

Bank's strategic approach

Members stressed the importance of a fully integrated agricultural approach, while ensuring cross-sectoral fertilization. They proposed integration of other sectors, particularly those associated with rural poverty: nutrition, health, and education and infrastructure and transport linked to market access. There was a need for a holistic approach to address systemic issues. Relatedly, a multifaceted approach was also needed given the complexity of this sector. A number of speakers requested maintaining consistency with the WDR, particularly its emphasis on the need to increase the productivity of smallholders and on improving governance.

Thematic performance

Some speakers felt the IEG study should have analyzed further the importance of gender, including recommendations on how to better align gender with the Bank's assistance. Several members stressed the importance of the Bank's engagement in land tenure and sustainable land management, while recognizing the political and social sensitivity of the matter and acknowledging that this is a country-driven process. Many speakers noted that the Bank has a role to play in promoting adaptation to climate change, responding to droughts, and improving infrastructure, including transport, roads, and water management. Relatedly, one member noted that the Bank does not have a comparative advantage in the agriculture processing industry and market-oriented products. The Bank should continue to scale up direct investment in irrigation, extension, and provision of fertilizers and improved seeds. Technology development and technology transfer for increasing productivity were also relevant. The need for new knowledge and institutional capacity, including for smallholders, as well as sustainability was also cited. There were also comments on the importance of farmers' access to credit, development of microfinance, and risk management instruments. In this vein, a speaker noted the importance of IFC's role in agricultural finance.

Aid architecture

Several speakers commented on the Bank's role and comparative advantage in the agriculture sector vis-à-vis other development partners. A few felt the Bank could play a coordination role in some areas such as donor financing or other commitments such as the implementation of the Paris Declaration on alignment and harmonization. One member noted the importance of aligning with the FAO and IFAD, which have recently conducted evaluations of their agriculture strategies. Others stressed the need for the Bank to integrate contributions from other entities such as CGIAR and national research centers. The high fragmentation of Bank and donor assistance in this sector was also mentioned.

Country focus

Some members stressed the importance of the demand-driven approach, based on a country's own prioritization. The Bank should help countries identify the binding constraints in the country-specific context. In addition, there was a need to address the role of the public sector vis-à-vis the private sector in the economy. Caution in promoting liberalization of the agriculture sector was requested. One member proposed making specific assessments based on the categorization of countries. In this regard, countries could be identified as predominantly agrarian societies, resource-rich, or relatively advanced, such as South Africa. This member felt the study could have benefited from views of country authorities, while noting that staff views were more on internal factors.

Analytic work

Some members stressed the need to focus on the Bank's comparative advantage in analytic work and policy advice. In this regard, some speakers felt that there are numerous studies, and the Bank should specialize in what it does well; that is, project management and monitoring and evaluation. One speaker felt the IEG study could have covered WBI activities—the nexus between research and implementation. *Management responded that analytical work under the agriculture pillar of the Africa Action Plan continues to be important and is being designed to be more strategic and quality-oriented. It also noted that monitoring and evaluation is integral to the Africa Action Plan, particularly in measuring productivity.*

Financing and staffing

Some speakers stressed the importance of ensuring appropriate human and financial resources, including through IDA, trust funds, and grants to address the agricultural challenges of the poorest African countries. One member regretted the progressive decline in the staff's technical skills, although welcoming management's efforts and recently undertaken organizational changes. A few members sought further clarification on the different number of technical experts presented by management and IEG. *IEG clarified that it drew on human resources data, which show that technical skills have declined since 1997. Management noted that 37 out of 79 staff (47 percent) were decentralized. Management also said that it was undertaking a comprehensive skill-mix review.*

Data

One member sought clarification on the divergence between management and IEG aggregated figures and country data. *IEG noted that while both IEG and management drew on World Development Indicators, management drew on aggregate growth rates, while IEG presented data by three categories of country performance.*

Jiayi Zou, Chairperson

Rapport de synthèse du Président : Comité pour l'efficacité du développement (CODE)

Le 3 octobre 2007, le Comité pour l'efficacité du développement (CODE) a examiné le rapport intitulé « *World Bank Assistance to Agriculture in Sub-Saharan Africa: An IEG Review* », préparé par le Groupe indépendant d'évaluation (IEG), ainsi que le projet de Réponse de la direction.

Rappel des faits

Le CODE a examiné l'évaluation de l'IEG intitulée « L'appui de la Banque mondiale au renforcement des capacités en Afrique » le 23 mars 2005. L'IEG a également préparé l'étude spéciale « *The World Bank's Assistance to Agricultural Water Management (1994–2004)* », datée du 7 juillet 2006.

Le projet de document « Rapport sur le développement dans le monde 2008 : l'agriculture au service du développement » a été examiné par le Conseil le 12 juin 2007 et doit être publié le 19 octobre. Son principal message est que l'agriculture reste un instrument de développement essentiel qui nécessite, en Afrique, qu'on lève les obstacles à l'augmentation de la productivité agricole des petits exploitants — qu'ils produisent essentiellement pour leur consommation ou qu'ils pratiquent déjà une agriculture commerciale.

Rapport de l'IEG

La première finalité de l'étude de l'IEG était de faire un premier bilan du recentrage des activités de la Banque sur l'agriculture en Afrique, tel qu'exprimé notamment dans le Plan d'action pour l'Afrique. Cet examen aidera par ailleurs à établir le schéma directeur de l'étude sur l'aide de la Banque mondiale à l'agriculture en général que l'IEG prévoit d'effectuer durant l'exercice 2009.

La présente étude évalue l'efficacité de l'aide apportée par la Banque mondiale pour remédier aux obstacles au développement agricole en Afrique subsaharienne pendant la période comprise entre les exercices 91 et 06. L'étude conclut que le secteur agricole a été négligé aussi bien par les autorités nationales que par la communauté des donateurs, Banque mondiale comprise. Elle constate que l'aide modérée de la Banque — en repli il y a peu encore — a été largement apportée au coup par coup et « saupoudrée » entre plusieurs domaines essentiels tels que la recherche, la vulgarisation, le crédit, les semences, les routes et les réformes, mais sans guère tirer parti des synergies existant entre eux. En conséquence, malgré la relative réussite affichée dans certains domaines, comme la recherche, les résultats sur le terrain ont été limités en raison du manque d'interaction avec la vulgarisation et de la disponibilité insuffisante de facteurs complémentaires critiques tels que les engrais, l'eau et l'accès aux marchés. De nombreux autres éléments entrent en ligne de compte pour expliquer les performances médiocres de l'agriculture en Afrique, à savoir la mauvaise gouvernance et les conflits qui sévissent dans plusieurs pays, les capacités institutionnelles insuffisantes, la sous-estimation par les États de l'importance de l'agriculture pour le développement, mais aussi le manque de coordination entre les donateurs.

L'étude note également qu'au fil du temps, les compétences techniques nécessaires à l'appui au développement agricole se sont amenuisées au sein de la Banque, et que ses travaux d'analyse ont été peu nombreux, de qualité variable et n'ont pas servi de fondement stratégique à l'élaboration des programmes de prêts et aux activités de conseil. L'étude recommande que la Banque : i) favorise l'augmentation de la productivité agricole en aidant à élaborer des mécanismes susceptibles de fournir aux agriculteurs tout un ensemble de facteurs, notamment des semences améliorées, de l'eau, des crédits et des services de vulgarisation de qualité, de façon coordonnée ; ii) s'emploie à renforcer ses propres capacités afin de pouvoir apporter une aide appropriée au développement de l'agriculture, en augmentant le nombre et la qualité de ses travaux d'analyse pour mieux définir les priorités dans chaque pays et en veillant à ce que ses activités de conseil et de prêt se fondent sur leurs conclusions, et en restaurant ses compétences techniques, et iii) perfectionne ses systèmes de données pour mieux suivre les opérations qu'elle finance, et consolide le dispositif de suivi et d'évaluation pour rendre correctement compte des activités conduites dans le cadre de projets dans différents pays.

Projet de Réponse de la direction

La direction souscrit dans leurs grandes lignes aux conclusions et recommandations formulées par l'IEG et a déjà commencé à les appliquer pour beaucoup d'entre elles. L'exercice stratégique engagé par la direction en vue d'identifier et de dimensionner des programmes agricoles intégrés sert de guide lors des tables rondes du CAADP et aide nos clients à élaborer leurs propres programmes. La direction ne partage pas l'avis de l'IEG sur l'importance donnée à certains aspects. Le rapport de l'IEG attribue le retard de croissance à des handicaps associés à la diversité agroécologique, à la pauvreté des sols, à la variabilité des précipitations et à la fréquence des sécheresses. La direction est d'accord avec ce constat, mais note également que les performances de l'agriculture sont affectées par certains paramètres techniques et par des facteurs économiques et institutionnels plus généraux. La direction estime que les recommandations de l'IEG abordent des points importants pour faire progresser l'agriculture en Afrique, mais que plusieurs d'entre elles (par exemple assurer un accès satisfaisant aux intrants) ont besoin d'être précisées avant de pouvoir être mises en œuvre. Pour la direction, il convient d'appuyer ces efforts de précision des recommandations dans le cadre de l'aide actuellement apportée aux pays pour leur permettre de définir des programmes intégrés. Par rapport à l'IEG, la direction estime qu'il convient de mettre davantage l'accent sur l'adhésion des pays aux programmes et sur l'harmonisation des actions avec les autres partenaires de développement.

Conclusions générales

Les membres du Comité ont apprécié l'examen de cette étude et les échanges de vues sur l'agriculture en Afrique qui viennent à point nommé, se félicitant en particulier que le président compte en faire l'un de ses principaux messages lors des Assemblées annuelles, et ont salué le regain d'attention porté au développement agricole pour favoriser la croissance économique et la réduction de la pauvreté. Les membres ont estimé que cet examen complétait bien le *Rapport sur le développement dans le monde 2008*, le Plan d'action pour l'Afrique et les actions menées en vue de la réalisation des ODM. Ils ont souligné la nécessité de veiller à ce que la Banque adopte une approche intégrée de la question agricole pour remédier aux problèmes systémiques, en repérant les obstacles les plus pénalisants dans chaque contexte national et en répondant aux besoins urgents des pays pauvres d'Afrique par une démarche plurisectorielle et multidimensionnelle. De l'avis général, des objectifs réalistes doivent être définis pour les interventions de la Banque, et un juste équilibre doit être trouvé entre les travaux d'analyse et les activités de conseil. Les Administrateurs se sont également déclarés favorables à l'idée de redynamiser l'intervention de la Banque en se fondant sur ses avantages comparatifs en matière de promotion du développement agricole, mais dans le cadre d'une coordination et d'une coopération claires avec d'autres partenaires de développement comme la FAO.

Les membres du Comité ont indiqué qu'il importait de prendre en compte l'avis des autorités nationales, de définir la place du service public par rapport au secteur privé, et de s'adapter à l'architecture de l'aide internationale et promouvoir la mise en œuvre de la Déclaration de Paris, en évitant la fragmentation de l'aide. Ils ont également souligné l'importance de s'attaquer aux problèmes de la gouvernance, de la corruption et des régimes fonciers, reconnaissant toutefois qu'il s'agissait là de questions sensibles du point de vue social, économique et politique. Dans le même esprit, tout en convenant de la nécessité d'obtenir l'adhésion des pays et de répondre à une demande, les membres ont exprimé des opinions diverses sur le rôle et l'implication de la Banque. Les membres du Comité ont aussi mis en avant les enjeux que représentent l'adaptation aux changements climatiques, l'accès au crédit pour les petits agriculteurs, et la mise au point d'instruments financiers novateurs, en particulier pour la gestion du risque, et l'importance d'avancer sur le dossier du commerce. Un certain nombre d'autres questions ont été évoquées par plusieurs intervenants, à savoir les moyens financiers nécessaires, l'utilisation de fonds fiduciaires et le rôle des prêts et des dons de l'IDA ; le développement de nouvelles technologies et le transfert de technologies ; la différenciation des pays en fonction de leur niveau de développement ; les capacités institutionnelles et la viabilité des réformes de la politique publique ; et des questions transversales telles que les infrastructures, les transports, l'eau, l'accès aux marchés et la promotion de la femme.

Prochaines étapes

Il a été demandé que la Stratégie de développement rural fasse l'objet d'un réexamen, le dernier bilan de l'exécution de la stratégie sectorielle ayant eu lieu en 2005 ainsi que l'a rappelé la direction. Les membres du Comité ont déclaré attendre avec intérêt l'étude sur l'aide de la Banque mondiale à l'agriculture en général que l'IEG prévoit d'effectuer durant l'exercice 09, bien que quelques intervenants aient demandé que ce rapport soit examiné plus tôt.

Les grands points ci-après ont été soulevés durant la réunion :

Le secteur agricole en Afrique

Les membres du Comité ont souligné l'importance du secteur agricole pour l'Afrique, sa contribution à la croissance, à la réduction de la pauvreté, et à la réalisation des objectifs de développement pour le Millénaire (ODM). Ils ont rejoint l'analyse de l'IEG que l'agriculture était largement négligée par les autorités nationales et les donateurs, et relevé que le volume des prêts à l'investissement accordés par la Banque au secteur avait été relativement modeste. Quelques membres ont fait observer que le problème de l'agriculture dépassait le cadre du continent africain. L'un d'entre eux a demandé un complément d'information sur l'encours des prêts. *La direction a répondu que le portefeuille de la Banque sur l'agriculture en Afrique comprenait 49 projets pour un montant de 2,2 milliards de dollars. Les nouveaux prêts, crédits et dons octroyés ces deux dernières années (exercices 06 et 07) ont dépassé les 550 millions de dollars chaque année, soit une augmentation de 80 % par rapport à la moyenne des exercices 01 à 05.*

Contribution de la Banque

Des déclarations ont été entendues sur la nécessité d'améliorer et étendre l'aide de la Banque à l'agriculture en Afrique, de lancer un appel d'offres pour rationaliser l'intervention de la Banque, et de préparer un rapport d'avancement sur les actions menées pour accroître la productivité agricole dans la Région. Un membre du Comité a insisté sur le fait qu'il importait d'aborder le dossier de l'agriculture africaine dans le contexte de l'Exercice stratégique à long terme engagé par le Groupe de la Banque mondiale sous la direction de son Économiste en chef. Dans le même ordre d'idées, le Groupe devrait maintenir ses efforts de coordination et de cohérence dans la programmation des activités pour chaque pays. Une question a été soulevée sur la nécessité éventuelle d'analyser les aspects organisationnels et les systèmes de gestion de la Banque.

Approche stratégique de la Banque

Les membres du Comité ont souligné l'importance d'adopter une approche totalement intégrée de l'agriculture, tout en assurant des échanges féconds entre les secteurs. Ils ont proposé l'intégration d'autres secteurs, en particulier ceux associés à la pauvreté rurale : la nutrition, la santé et l'éducation, et les infrastructures et les transports liés à l'accès aux marchés. Une approche globale était nécessaire pour remédier aux problèmes systémiques. De même, une approche multidimensionnelle était également indispensable compte tenu de la complexité de ce secteur. Un certain nombre d'intervenants ont demandé que l'on reste dans la ligne du RDM, et notamment de l'accent qui y est mis sur le besoin d'augmenter la productivité des petits agriculteurs et sur l'amélioration de la gouvernance.

Résultats par domaines

Certains intervenants ont estimé que l'étude de l'IEG aurait dû analyser davantage l'importance de la promotion de la femme, et comporter notamment des recommandations sur la manière de mieux intégrer cet aspect dans l'aide de la Banque. Plusieurs membres ont souligné l'importance de l'intervention de la Banque dans le domaine des régimes fonciers et de la gestion durable des terres, tout en reconnaissant le caractère politiquement et socialement sensible du problème, auquel il revient aux pays de s'attaquer. De nombreux intervenants ont noté que la Banque avait un rôle à jouer pour ce qui était d'encourager les pays à s'adapter aux changements climatiques, de faire face aux sécheresses et d'améliorer les infrastructures, notamment dans les transports, le réseau routier et la gestion de l'eau. À ce sujet, un membre a fait remarquer que la Banque ne possédait pas d'avantage comparatif dans l'industrie agroalimentaire ni dans les produits « orientés marché ». La Banque devait continuer à accroître ses investissements directs dans l'irrigation, la vulgarisation et la fourniture d'engrais et de semences améliorées. Le développement de nouvelles technologies et le transfert de technologies permettant d'accroître la productivité étaient également importants. La nécessité de renforcer les savoirs et les capacités institutionnelles, y compris pour les petits exploitants, ainsi que la viabilité, a aussi été mentionnée. Des remarques ont été formulées par ailleurs sur l'importance de l'accès des agriculteurs au crédit, du développement du microcrédit, et des instruments de gestion du risque. À cet égard, un intervenant a noté le rôle majeur de l'IFC dans la finance agricole.

Architecture de l'aide

Plusieurs intervenants ont évoqué le rôle de la Banque et son avantage comparatif dans le secteur agricole par rapport aux autres partenaires de développement. Quelques-uns ont estimé que la Banque pouvait jouer un rôle de coordination dans certains domaines, par exemple les financements accordés par les donateurs, ou d'autres engagements tels que la mise en œuvre de la Déclaration de Paris sur l'alignement et l'harmonisation de l'aide. Un membre du Comité a signalé l'importance de travailler en phase avec la FAO et le FIDA, qui ont procédé récemment à des évaluations de leurs stratégies pour l'agriculture. D'autres ont souligné la nécessité pour la Banque d'intégrer les contributions d'autres organismes comme le CGIAR et les centres de recherche nationaux. Il a aussi été fait référence à la fragmentation importante de l'aide de la Banque et des autres donateurs dans ce secteur.

Spécificités nationales

Certains membres du Comité ont insisté sur l'importance d'intervenir en réponse à une demande, en fonction des priorités définies par chaque pays. La Banque devrait aider les pays à repérer les obstacles les plus pénalisants dans leur contexte national. Il fallait également repenser la place du secteur public par rapport au secteur privé dans l'économie. Une certaine prudence a été demandée pour ce qui est de promouvoir la libéralisation de l'agriculture. Un membre a proposé de réaliser des évaluations en différenciant plusieurs catégories de pays. On pouvait ainsi caractériser les pays comme étant des sociétés à dominante agraire, riches en ressources ou relativement avancées, comme l'Afrique du Sud. Ce membre a jugé que l'étude aurait pu gagner à présenter les opinions des autorités nationales, notant toutefois que les avis des services de la Banque portaient davantage sur les facteurs internes.

Travaux d'analyse

Certains membres ont souligné la nécessité d'axer les travaux d'analyse et les activités de conseil sur l'avantage comparatif de la Banque. À cet égard, des intervenants ont estimé qu'il existait de nombreuses études, et que la Banque devrait se spécialiser dans ce qu'elle sait faire bien, à savoir la gestion de projet, et le suivi et l'évaluation. Un intervenant a indiqué que l'étude de l'IEG aurait pu couvrir les activités du WBI, qui fait le lien entre recherche et application. *La direction a répondu que les travaux d'analyse sur l'agriculture entrepris dans le cadre du Plan d'action pour l'Afrique restaient une tâche importante à laquelle on s'employait à donner une orientation plus stratégique et plus qualitative. Elle a fait remarquer également que le suivi et l'évaluation faisaient partie intégrante du Plan d'action pour l'Afrique, en particulier pour ce qui est de la mesure de la productivité.*

Financement et personnel

Certains intervenants ont insisté sur l'importance de disposer de ressources humaines et financières suffisantes, y compris à travers l'IDA, les fonds fiduciaires et les dons, pour s'attaquer aux problèmes de l'agriculture dans les pays africains les plus pauvres. Un membre du Comité a regretté le déclin progressif des compétences techniques du personnel, saluant néanmoins les efforts faits par la direction et la réorganisation entreprise récemment. Quelques membres ont demandé des éclaircissements sur la différence entre le nombre d'experts techniques indiqué par la direction et celui donné par l'IEG. *L'IEG a expliqué qu'il utilisait les données du service des ressources humaines, qui montrent que les compétences techniques ont diminué depuis 1997. La direction a fait observer que sur les 79 membres du personnel, 37 (47 %) étaient décentralisés. Elle a indiqué également qu'elle était en train d'entreprendre un état des lieux complet des compétences.*

Données

Un membre du Comité a demandé des éclaircissements sur l'écart entre les chiffres totaux et les données sur les pays fournis par la direction et par l'IEG. *L'IEG a déclaré que l'IEG et la direction tiraient tous les deux leurs chiffres des Indicateurs du développement dans le monde, mais que la direction utilisait les taux de croissance globaux, tandis que l'IEG présentait des données selon trois catégories de pays classés en fonction de leurs performances.*

Jiayi Zou, Président

Resumo do Presidente: Comissão sobre a Eficácia do Desenvolvimento (COED)

Em 3 de Outubro de 2007 a Comissão sobre a Eficácia do Desenvolvimento (COED) examinou o relatório intitulado *A Assistência do Banco Mundial à África Subsariana: A Análise do GIA*, elaborado pelo Grupo Independente de Avaliação (GIA), juntamente com a Resposta Preliminar da Administração.

Antecedentes

A COED examinou a avaliação do GIA com o título de *O Apoio do Banco Mundial para o Aumento das Capacidades na África* em 23 de Março de 2005. O GIA também elaborou um estudo especial intitulado *A Assistência do Banco Mundial à Gestão dos Recursos Hídricos na Agricultura* (1994 – 2004), com data de 7 de Julho de 2006.

A versão preliminar do *Relatório sobre o Desenvolvimento Mundial de 2008: A Agricultura para o Desenvolvimento* foi objecto das deliberações do Conselho de Administração do Banco em 12 de Junho de 2007, devendo o relatório ser publicado no dia 19 de Outubro. A principal mensagem nele contida é que a agricultura continua a ser um instrumento fundamental para o desenvolvimento, e que no caso da África isto requer que sejam abordados os constrangimentos à produtividade agrícola dos pequenos agricultores— tanto aqueles que dela dependem para a sua subsistência como aqueles que exercem uma actividade comercial.

O Relatório do GIA

A análise do GIA destinava-se principalmente a proporcionar uma visão e avaliação, em tempo oportuno, da focalização renovada do Banco na agricultura da África, especialmente tal como incorporada no Plano de Acção para a África. Ela também vai ajudar a definir a estrutura do projecto de estudo do GIA sobre a assistência à agricultura prestada por todo o Banco, programada para o ano fiscal de 2009.

O estudo avalia a eficácia para o desenvolvimento da assistência do Banco Mundial (BM) para abordar os constrangimentos existentes ao desenvolvimento da agricultura na África Subsariana durante o período compreendido entre os anos fiscais de 1991 e 2006. O estudo faz notar que o sector da agricultura tem sido negligenciado tanto pelos governos como pela comunidade de doadores, incluindo o Banco Mundial. Ele verifica que o apoio limitado e, até recentemente, decrescente do Banco tem sido em grande medida prestado aos poucos e por assim dizer "salpicado" através de várias áreas essenciais, tais como, a investigação, a extensão, o crédito, as sementes, as estradas e as reformas de política, porém com um reconhecimento escasso das sinergias existentes entre essas áreas. Em consequência, embora algumas áreas tenham tido um êxito comparativo maior — por exemplo, a investigação — os resultados no terreno foram limitados devido aos vínculos débeis com a extensão a à disponibilidade limitada de factores complementares essenciais, tais como, os fertilizantes, a água e o acesso aos mercados. A governação deficiente e os conflitos em diversos países, a debilidade das capacidades institucionais, e a apreciação inadequada dos governos da importância da agricultura no desenvolvimento, assim como a coordenação insuficiente dos esforços dos doadores, têm contribuído para o fraco desempenho da agricultura no continente.

O estudo também verifica que as competências técnicas do Banco para apoiar o desenvolvimento da agricultura diminuíram com o tempo, e que o seu trabalho analítico tem sido limitado, de qualidade variável, e que não tem informado estrategicamente a concepção dos programas de empréstimos ou o assessoramento de política. O estudo recomenda que o Banco: (i) Apoie o melhoramento da produtividade agrícola ajudando a conceber mecanismos que possam trazer vários factores, tais como, sementes melhoradas, água, crédito e bons conselhos sobre a extensão, entre outros, aos agricultores, de uma maneira coordenada; (ii) Se concentre em melhorar as suas próprias capacidades para apoiar adequadamente ao desenvolvimento da agricultura, aumentando a quantidade e a qualidade do trabalho analítico para ajudar a estabelecer as prioridades a nível dos países e a assegurar que a assessoria de política e os empréstimos se fundem nas suas conclusões, e reedificando as suas competências técnicas; (iii) Melhore os seus sistemas de dados para melhor fazer o seguimento das actividades apoiadas pelo Banco, e reforçar o monitoramento e a avaliação (M&A) para informarem com exactidão sobre as actividades dos projectos nos vários países.

Resposta Preliminar da Administração

A Administração concorda, no sentido amplo, com as conclusões e recomendações do GIA e já está a por muitas delas em prática. O exercício estratégico em curso realizado pela Administração, para identificar e definir as dimensões de programas abrangentes para a agricultura, informa as Mesas Redondas do Programa Abrangente para o Desenvolvimento da Agricultura na África (PADAA - CAADP) e ajuda os seus clientes a definir os seus próprios programas. A Administração difere do GIA em algumas áreas de destaque. A análise do GIA atribui o lento crescimento a constrangimentos associados à diversidade agro-ecológica, à pobreza dos solos, à pluviosidade variável, e às secas frequentes. A Administração concorda, mas também assinala que factores, tanto económicos como institucionais, mais amplos afectam o desempenho. A Administração acha que as recomendações do GIA abordam questões que são importantes para o avanço da agricultura na África, mas que várias das recomendações específicas (por exemplo, assegurar o acesso aos insumos em tempo oportuno) têm que ser mais detalhadas para que possam ser postas em prática. A Administração está a apoiar maiores precisões que são necessárias, no âmbito da assistência já está prestada aos países, para definir programas abrangentes. A Administração difere do GIA no que respeita a dar mais peso ao envolvimento dos países nos programas e à harmonização com os outros parceiros no desenvolvimento.

Conclusões Globais

Os membros saudaram o debate oportuno sobre este estudo e sobre a agricultura na África, especialmente a sua inclusão prevista como mensagem principal do presidente para as Reuniões Anuais, e o destaque renovado dado ao desenvolvimento da agricultura para o crescimento económico e a redução da pobreza. Os membros pensaram que o debate era um bom complemento do *Relatório sobre o Desenvolvimento Mundial de 2008* (RDM), do Plano de Acção para a África e do trabalho no sentido de alcançar os OMD. Os membros frisaram a necessidade de assegurar que o Banco siga uma abordagem integrada à agricultura para que as questões sistémicas sejam abordadas, identificando os constrangimentos que existem obrigatoriamente no contexto dos diferentes países, e tratar das necessidades urgentes dos países pobres africanos mediante uma abordagem multissectorial e multifacetada. Houve consenso sobre a necessidade de estabelecer metas realistas para a participação do Banco e um equilíbrio apropriado entre o trabalho analítico e o assessoramento de política. Os Administradores também apoiaram os esforços tendentes a revigorar o comprometimento do Banco com base na sua vantagem comparativa para promover o desenvolvimento da agricultura, mas evidentemente com a coordenação e cooperação de outros parceiros no desenvolvimento, tais como a FAO.

Os membros teceram comentários sobre a necessidade de incluir as opiniões dos governos, de definir o papel do sector público em relação ao sector privado, e de se adaptar à arquitectura mundial das ajudas e avançar na implementação da Declaração de Paris, evitando a fragmentação da assistência. Também frisaram a importância de abordar as questões relacionadas com a governação, a corrupção e a propriedade das terras, reconhecendo porém as sensibilidades sociais, económicas e políticas deste assunto. Nesta ordem de ideias, se bem que estejam de acordo com a necessidade de seguir uma abordagem de envolvimento dos países e induzida pela procura, os membros expressaram opiniões diversas sobre a função e a participação do Banco. Os membros também sublinharam a adaptação às alterações climáticas, o acesso ao crédito pelos pequenos agricultores, e a criação de instrumentos financeiros inovadores, especialmente para a gestão dos riscos, assim como a importância de abordar a agenda do comércio. Outros assuntos que causaram uma impressão positiva a vários participantes foram: os recursos financeiros necessários, a utilização de fundos fiduciários, e a função dos empréstimos e doações da AID; o desenvolvimento da tecnologia e a transferência de tecnologias; a categorização dos países levando em conta os níveis específicos de desenvolvimento; as capacidades das instituições e a sustentabilidade das reformas de política; e assuntos comuns a vários sectores, tais como, as infra-estruturas, os transportes, a água, o acesso aos mercados e a dimensão de género.

Os Próximos Passos a Dar

Foi recebido um pedido de actualização da Estratégia Sectorial para o Desenvolvimento Rural, a qual, segundo apontado pela Administração, tinha sido debatida pela última vez na Actualização da Execução da Estratégia Sectorial de 2005. Os membros aguardavam com interesse o exame que seria feito do estudo proposto do GIA sobre a assistência a ser prestada pelo Banco, na sua totalidade, à agricultura no ano fiscal de 2009, embora alguns oradores tenham pedido que o relatório fosse examinado mais cedo.

Os seguintes assuntos principais foram levantados durante a reunião:

O sector da agricultura na África

Os membros frisaram a importância do sector da agricultura para a África, a sua contribuição para o crescimento e a redução da pobreza, e para alcançar os Objectivos do Milénio para o Desenvolvimento (ODM). Concordaram com o GIA que a agricultura era, em grande medida, negligenciada pelos governos e doadores, e assinalaram que os empréstimos para o desenvolvimento concedidos pelo Banco a este sector tinham sido relativamente poucos. Alguns membros fizeram notar que o problema da agricultura existe para além da Região da África. Um deles procurou obter mais informações sobre o volume dos empréstimos por liquidar. *A Administração respondeu que a carteira de 49 projectos para a agricultura na África é de USD 2.2 biliões. Os novos empréstimos, créditos e doações relativos aos últimos dois anos (anos fiscais de 2006 e 2007) elevaram-se a mais de USD 550 milhões em cada ano, ou seja, um aumento de 80 por cento comparado com a média dos anos fiscais de 2001 a 2005.*

A contribuição do Banco

Foram feitos comentários sobre a necessidade de o Banco reforçar e aumentar o apoio que presta à agricultura na África, um pedido de uma proposta relativa à racionalização da participação do Banco, e um relatório sobre a situação das actividades destinadas a incrementar a produtividade agrícola na Região. Um membro sublinhou a necessidade de considerar a agricultura da África no contexto do Exercício para uma Estratégia a Longo Prazo do Grupo do Banco Mundial, chefiado pelo economista principal. De maneira conexa, o Grupo do Banco deveria manter a coordenação e compatibilidade entre os programas dos países. Foi feita a pergunta de saber se havia necessidade de examinar os aspectos organizativos e os sistemas de gestão do Banco.

A abordagem estratégica do Banco

Os membros sublinharam a importância de seguir uma abordagem totalmente integrada à agricultura, assegurando porém uma fertilização Multissectorial. Propuseram a integração de outros sectores, especialmente daqueles que estão associados à pobreza rural: nutrição, saúde e educação, e infra-estruturas e transportes ligados ao acesso aos mercados. Era necessário seguir uma abordagem holística para tratar das questões sistémicas. De maneira conexa, era também necessário seguir uma abordagem multifacetada devido à complexidade deste sector. Alguns oradores pediram que se mantivesse a coerência com o RDM, em especial no que respeita destacar a necessidade de aumentar a produtividade dos pequenos produtores e de uma melhor governação.

Desempenho temático

Alguns oradores pensavam que o estudo do GIA deveria ter analisado mais a importância do género, incluindo recomendações sobre a melhor maneira de alinhar as questões de género à assistência prestada pelo Banco. Diversos membros destacaram a importância do comprometimento do Banco com a propriedade das terras e a gestão sustentável das mesmas, se bem que reconhecendo a sensibilidade política e social do assunto e reconhecendo que esse é um processo induzido pelos países. Muitos oradores fizeram notar que o Banco tem um papel a desempenhar na promoção da adaptação às alterações climáticas, na reacção às secas, e no melhoramento das infra-estruturas, incluindo os transportes, as estradas e a gestão dos recursos hídricos. De maneira conexa, um membro assinalou que o Banco não tem uma vantagem comparativa na indústria transformadora dos produtos agrícolas e dos produtos orientados para o mercado. O Banco deveria continuar a aumentar o investimento directo na irrigação, na extensão e no fornecimento de fertilizantes e sementes melhoradas. O desenvolvimento de tecnologias e a transferência de tecnologias para aumentar a produtividade também eram importantes. Foi também citada a necessidade de novos conhecimentos e capacidades institucionais, inclusivamente para os pequenos agricultores, assim como a sustentabilidade. Foram igualmente feitos comentários sobre a importância do acesso ao crédito pelos agricultores, do desenvolvimento de micro financiamento e de instrumentos para a gestão dos riscos. Nesta mesma ordem de ideias, um orador referiu-se à importância da função desempenhada pela SFI no financiamento da agricultura.

A arquitectura das ajudas

Diversos oradores comentaram sobre o papel do Banco e a sua vantagem comparativa no sector da agricultura em relação aos outros parceiros que obram em prol do desenvolvimento. Alguns pensavam que o Banco podia exercer uma função de coordenador em algumas áreas, tais como, o financiamento dos doadores ou outros comprometimentos como a aplicação da Declaração de Paris relativamente ao alinhamento e à harmonização. Um membro mencionou a importância do alinhamento com a Organização das Nações Unidas para a Agricultura e a Alimentação (FAO) e o Fundo Internacional de Desenvolvimento Agrícola (FIDA), os quais realizaram recentemente uma avaliação das suas estratégias no campo da agricultura. Outros insistiram sobre a necessidade de o Banco integrar as contribuições de outras entidades, tais como, o CGIAR e os centros nacionais de investigação. A grande fragmentação da assistência do Banco e dos doadores neste sector também foi mencionada.

Focalização nos países

Alguns membros frisaram a importância de uma abordagem induzida pela procura, baseada nas prioridades estabelecidas pelos próprios países. O Banco deveria ajudar os países a identificar os constrangimentos obrigatórios no contexto específico dos países. Adicionalmente, era necessário abordar o papel desempenhado pelo sector público face ao sector privado na economia. Solicitou-se cautela na promoção da liberalização do sector da agricultura. Um membro propôs que fossem feitas avaliações específicas baseadas na categorização dos países. A este respeito, os países poderiam ser identificados como sociedades predominantemente agrárias, ricas em recursos ou relativamente avançadas, tais como a África do Sul. Este membro era de opinião que o estudo poderia ter beneficiado das opiniões das autoridades dos países, fazendo porém notar que as opiniões dos funcionários do Banco incidiam mais sobre factores internos.

Trabalho analítico

Alguns membros frisaram a necessidade de focalizar a vantagem comparativa do Banco no trabalho analítico e no assessoramento de política. A este respeito, alguns oradores pensavam que existem numerosos estudos, e que o Banco deveria especializar-se naquilo que faz bem, isto é, na gestão, no monitoramento e na avaliação dos projectos. Um orador pensava que o estudo do GIA poderia ter coberto as actividades do Instituto do Banco Mundial (IBM) — o nexo entre a investigação e a implementação. *A Administração respondeu que o trabalho analítico sob o pilar do Plano de Acção para a África continua a ser importante e está a ser concebido de forma a ser mais estratégico e orientado para a qualidade. Também fez notar que o monitoramento e a avaliação fazem parte integrante do Plano de Acção para a África, especialmente no que respeita a medição da produtividade.*

Financiamento e dotação de pessoal

Alguns oradores sublinharam a importância de assegurar recursos humanos e financeiros apropriados, inclusivamente através da AID, de fundos fiduciários, e doações para lidar com os desafios que se colocam à agricultura nos países mais pobres da África. Um membro lamentou o declínio progressivo das competências técnicas dos funcionários, saudando ao mesmo tempo os esforços da Administração e as alterações organizativas recentemente introduzidas. Alguns membros procuraram obter esclarecimentos adicionais sobre o número diferente de especialistas técnicos apresentado pela Administração e pelo GIA. *O GIA esclareceu que recorreu a dados relativos aos recursos humanos, os quais mostram que as competências técnicas vêm diminuindo desde 1997. A Administração fez notar que 37 entre 79 funcionários (47 por cento) foram descentralizados. A Administração referiu também que estava a empreender uma análise abrangente da composição das competências.*

Dados

Um membro pediu um esclarecimento sobre a divergência existente entre os números agregados apresentados pela Administração e pelo GIA e entre os dados sobre os países. *O GIA esclareceu que, se bem que tanto o GIA como a Administração se tenham baseado no Indicadores do Desenvolvimento Mundial, a Administração tomou as taxas de crescimento agregadas enquanto que o GIA apresentou dados em três categorias do desempenho dos países.*

Jiayi Zou, Presidente

Women harvesting cotton in Madagascar. Photo by Yosef Hadar, courtesy of World Bank Photo Library.

Evaluation Snapshot in Selected Languages

English

- Underperformance of agriculture has been a major limitation of Africa's development. For most of the past two decades, both governments and donors, including the World Bank, have neglected the sector.
- The Bank's limited—and, until recently, declining—support to agriculture has not been strategically used to meet the diverse needs of a sector that requires coordinated interventions across a range of activities. Lending from the Bank has been sprinkled across various agricultural activities such as research, extension, credit, seeds, and policy reforms in rural space, but with insufficient recognition of the synergies among them.
- The Bank now has an opportunity, drawing on its comparative advantage as a multisector lending institution and as the single largest donor to African agriculture (during 1990–2005), to help ensure a coordinated and multifaceted approach to agriculture development in Africa.

French — Français

- Les mauvais résultats de l'agriculture ont été un frein essentiel au développement de l'Afrique. Les autorités nationales et les bailleurs de fonds, dont la Banque mondiale, ont négligé ce secteur pendant l'essentiel des vingt années écoulées.
- L'aide modérée et, jusque récemment, en recul de la Banque mondiale à l'agriculture n'a pas été utilisée de manière stratégique pour répondre aux besoins variés d'un secteur qui nécessite des interventions coordonnées dans des domaines très divers. Ses crédits ont été dispersés entre différentes activités agricoles telles que la recherche, la vulgarisation, le crédit, les semences et les réformes de l'espace rural, mais sans tenir suffisamment compte de leurs synergies.
- La Banque mondiale a aujourd'hui la possibilité, en s'appuyant sur l'avantage comparatif qu'elle détient en tant qu'institution de crédit multisectorielle et premier bailleur de fonds à l'agriculture africaine (de 1990 à 2005), d'adopter une approche coordonnée et pluridimensionnelle au développement de l'agriculture sur ce continent.

Portuguese **Português**

- O fraco desempenho da agricultura tem sido uma limitação importante para o desenvolvimento da África. Durante a maior parte das duas últimas décadas, tanto o governo como os doadores, incluindo o Banco Mundial, negligenciaram este sector.
- O apoio limitado, e até recentemente decrescente, do Banco à agricultura não foi utilizado estrategicamente para suprir as diversas necessidades de um sector que requer intervenções coordenadas através de várias actividades. Os empréstimos do Banco têm sido espalhados por várias actividades agrícolas, tais como, a investigação, a extensão, o crédito, as sementes e as reformas de política no espaço rural, mas não houve um reconhecimento suficiente das sinergias existentes entre elas.
- O Banco tem agora a oportunidade, aproveitando a sua vantagem comparativa como instituição de crédito multilateral e sendo o maior doador para a agricultura africana (durante o período compreendido entre 1990 e 2005), de assegurar que seja seguida uma abordagem coordenada e multifacetada ao desenvolvimento da agricultura na África.

Chapter 1

Evaluation Highlights

- Sub-Saharan Africa is a diverse and complex Region and is behind on most of the Millennium Development Goals.
- Agricultural development can make a major contribution to poverty alleviation and growth.
- Increasing agricultural productivity is key to improved food security for both rural and urban poor.

Picking tomatoes in irrigated fields, Senegal River Basin. Photo by Scott Wallace, courtesy of World Bank Photo Library.

Introduction

Sub-Saharan Africa is a diverse and complex Region with more than 700 million people and at least 1,000 different ethnic groups in 47 countries with 7 distinctly different colonial histories. Some of the world's poorest countries are in the Region, and during the past two decades, the number of Africa's poor has doubled, from 150 million to 300 million, constituting more than 40 percent of the Region's population (World Bank 2005e).

According to the *World Development Report 2008,* the rural poverty rate in the Region was 82 percent in 2002.[1] Africa remains behind on most of the Millennium Development Goals (MDGs). As a result, the Region and its development are now a priority for the international community.

A major reason that Africa lags behind other Regions is the underperformance of its agriculture, which accounts for 30 percent of the gross domestic product (GDP) and employs 75 percent of the population (Commission for Africa 2005). The weak performance of the sector is the result of a variety of constraints that are particular to agriculture in Africa and make its development a complex challenge. Poor governance and conflict in several countries makes things even more difficult.

Total agricultural output in Africa consists primarily of food crops; agricultural export crops account for only 8 percent of total agricultural production (Peacock, Ward, and Gambarelli 2007). While some export crops, such as cotton, have often been considered an African success story (see appendix I), food crops have performed particularly poorly in most countries. Cereal yields in Africa, even in 2003–05, were less than half those in South Asia and one-third those in Latin America. Africa also lags behind other Regions in percentage of cropland irrigated, fertilizer use, and land and labor productivity per worker (table B4, appendix B). The underperformance of the sector initially led to skepticism about agriculture's potential to contribute to Africa's growth and poverty reduction (Diao and others 2006). But the weak performance of Africa's agriculture is attributable to a variety of factors that are unique to the sector in that Region. This evaluation of the World Bank's contribution to development of the agriculture sector in Sub-Saharan Africa provides some insights into these reasons based on Bank experience.

The Role of Agriculture in Africa

If Africa is to achieve the MDGs, its agriculture sector has to be developed. Until recently the sector was neglected because neither governments nor donors made its development a priority. In the immediate post-independence era, during the 1960s, governments in several African countries treated agriculture primarily as a source of resources for industrialization, in the belief that industrialization was the way to development and food aid could meet the needs of cities and help deal with emergencies.[2] Production of cash crops was encouraged as a source of foreign exchange for development.

Achievement of the MDGs in Africa will require realization of agriculture's potential.

Then, in the 1970s, World Bank President Robert McNamara led the shift from an economic growth paradigm to a broader development paradigm in Africa. This committed the Bank to integrated rural development to directly attack Africa's rural poverty and underdevelopment (Eicher 1999). While in Asia this broader rural focus came after the initial focus on food production and the building of institutions, serious focus on agricultural development by donors did not take place in Africa because of this shift in priorities.[3]

Agriculture has strong indirect effects on growth in other sectors.

Later, when African countries were faced with severe fiscal crises in the mid-1980s, donors prioritized improvements in the efficiency of resource allocation. In the agriculture sector, more emphasis was given to marketing reforms, rather than to the development of all relevant activities in the sector. Success with marketing reforms was considered a crucial determinant of the overall response of the economy to changing economic incentives.

Moving forward, a focus on agricultural development is critical to contribute to poverty reduction and economic growth in the Region.

Poverty reduction

Farming in Africa is largely a household enterprise, and most farmers have 0.5 to 2.0 hectares of land. For most of them, the small piece of land they farm (whether or not they own it) is their only tangible asset. This differs sharply from the situation in South Asia, where most of the poor are landless (Lipton and others 2003).[4] Low productivity and not landlessness is the major problem in Africa. Under such circumstances, increasing the productivity of small pieces of land has the potential to reduce poverty significantly in the Region.

Farming in Africa is largely carried out by smallholder farmers, and low productivity and not landlessness is the major problem.

The relationship between poverty reduction and agriculture in Africa is a powerful one. However, it is not always sufficiently appreciated that productivity improvement not only increases the food security of the rural poor, but also benefits the urban poor, for whom increased production means lower food prices.[5] Based on work in eight countries in the Region, Dorosh and Haggblade (2003) found that investments in agriculture generally favor Africa's poor more than similar investments in manufacturing.[6] IFPRI research (2002b) shows that each 10 percent increase in smallholder agricultural productivity in Africa can move almost 7 million people above the dollar-a-day poverty line. Recent Bank analytical work has found similar favorable results for poverty reduction arising from increased agricultural production (World Bank 2005j). Hartmann (2004) has gone as far as to note that if the development community had to choose just one activity with which to address the first MDG of reducing extreme poverty and hunger in Africa, it should be to produce more food.

Increased agricultural productivity improves food security for both rural and urban poor.

Growth

Recent research demonstrates that the effect of agriculture on wider growth is also likely to be substantial. Christiaensen and Demery (2007) distinguish between the direct and indirect effects of this growth and argue that while agriculture tends to grow more slowly than non-agriculture, the indirect effects of agriculture on non-agriculture are substantially larger than the reverse feedback effects. These effects arise from linkages to agro-processing and input production, for example, as well as from the "wage good effect," which means that lower food prices imply an increase in saving at a given level of income and can stimulate demand for goods produced by the non-agriculture sector.

Study Purpose

The purpose of this IEG review is twofold. First, it serves as a pilot for the proposed IEG study of Bank-wide assistance in agriculture scheduled for fiscal 2009. Second, the review provides timely insight into specific issues relevant to the Bank's renewed focus on agriculture in Africa, especially as expressed in the Africa Action Plan. In addition, the African Union has launched a vision and strategic framework for Africa's renewal—the New Partnership for Africa's Development (NEPAD). The Comprehensive Africa Agriculture Development Programme is at the heart of efforts by African governments under the NEPAD initiative to accelerate growth and eliminate poverty and hunger. Lessons of experience from the Bank will contribute to the discussion surrounding these initiatives and will likely inform future international aid agendas and policy directions.

Study Scope

The focus of the study is agricultural development, not the broader issue of rural development, in Africa over the 15-year period of 1991–2006. The 47 countries of the Region are highly diverse in resources, endowments (see table B.1, appendix B for categorization), and ability to commit politically to actions that increase growth and reduce poverty (World Bank 2002a). Given this diversity, the study focuses primarily on the common issues across countries that are relevant for agricultural development in the Region as derived largely from a limited set of strategic statements of the Bank. The scope of the review is also influenced by the following:

- Only the Bank's direct lending and nonlending activities have been considered as a part of this study. The Bank was the single largest donor to African agriculture during 1990–2005, and an evaluation of its activities can provide valuable insights into the challenge of agricultural development in the Region. Undoubtedly greater value could be added if Bank support could be assessed along with the activities supported by other donors. However, such an exercise cannot be easily carried out with modest resources. Other multilateral organizations, such as the Food and Agriculture Organization (FAO), African Development Bank (AfDB), and the International Fund for Agricultural Development (IFAD) are also currently evaluating their support to the agriculture sector in Africa. When these evaluations are complete, they will be brought to bear in IEG's forthcoming study of Bank-wide assistance to agriculture.
- The report draws on the findings of other IEG studies that have reviewed regional and global programs in support of agricultural development, but did not itself encompass a comprehensive review of all related regional and global programs.
- Although food security is discussed, the report does not discuss the merits or demerits of food aid.
- The discussion on market access for agricultural products is confined to transport infrastructure and does not extend to other barriers, such as those arising from the need for conformity with specifications demanded by supermarkets.

Methodology

The evaluation is built on four main sources of information:

- ***Portfolio review:*** In consultation with the Bank's Agriculture and Rural Development (ARD) Department, IEG identified a portfolio of projects with agriculture components for review. Trends in lending for the portfolio of 262 projects were examined. In addition, a stratified random sample of 71 closed and ongoing projects was selected from the portfolio for detailed review. The Bank's nonlending activities (including relevant rural strategy documents), Country Assistance Strategies, and Poverty Reduction Strategy Papers were also examined to assess the Bank's strategic approach to the development of the agriculture sector.
- ***Country-level reviews:*** Two countries in East Africa (Kenya and Malawi) and two in West Africa (Cameroon and Nigeria) were selected for sector reviews to provide country-specific insights. Assessments of 13 agricultural projects in various African countries were also fielded by IEG in fiscal 2007.
- ***Literature review:*** Bank and non-Bank literature provided a basis for understanding the complexities in African agriculture and the Bank's role, as well as for confirming the findings of the portfolio analysis and the country-level reviews.
- ***IEG Bank staff survey:*** Bank staff (both headquarters and field-based) views on internal factors and incentives related to the Bank's assistance were sought. The survey was sent to 258 staff who worked on agricultural issues in the Africa Region and in the ARD Network as agriculture specialists or as task managers of projects with agricultural components, in-

cluding projects in sectors such as transport and multisector operations.

Some limitations of study design

The study has two main limitations. First, although project assessments provide the field input and bring the perspectives of government officials and other stakeholders on the Bank's support to agriculture, the study is largely a desk review carried out over 8 months and on a limited budget (compared with typical IEG sector/thematic studies). Second, the response rate of the staff survey was only 22 percent. Since it is in the nature of opinion surveys to have a response bias, it is difficult to ascertain whether those who responded are representative of the 258 staff to whom the survey was originally sent. Because of the limited number of responses and the likely response bias, the survey results have been used only to illustrate and/or substantiate the findings from other information sources. Details on the methodology are included in appendix A.

Chapter 2

Evaluation Highlights

- Because of Africa's agro-ecological diversity, climate variability, poor soils, and limited irrigation, development of African agriculture is a complex challenge.
- The strategy for agricultural development in Africa will need to be based on a recognition of the Region's particular characteristics.
- If improved seeds, water, infrastructure, credit, and extension, among other measures, are made available at the same time or in optimal sequence, rapid growth in agricultural incomes is achievable in Africa.

Minibus piled high with goods and animals, Burkina Faso. Photo by Curt Carnemark, courtesy of World Bank Photo Library.

African Agriculture and the Bank

This section provides a brief background to the agriculture sector in Sub-Saharan Africa, followed by an examination of the Bank's strategic approach for development of the sector. It also identifies the main constraints to agricultural development in the Region.

The Agriculture Sector in Africa

Agriculture in Africa is primarily a private family activity, carried out largely by smallholders. Women provide about half of the labor force and produce most of the food crops consumed by the family. In some countries women's share in agricultural labor is even larger. In the Republic of Congo, for example, 70 percent of those involved in food crop production are women. While agricultural output is growing in Africa, labor productivity in the sector has been low and stagnant over most of the past two decades (World Bank 2002a).

Calculating a reliable growth rate for African agriculture over the study period is difficult because of the wide variation among countries and over time. The 47 countries in the Region can be divided into three categories: the comparatively better performers, with agricultural growth above 5 percent per year during 2000–04; the medium performers, with agricultural growth between 2 and 5 percent; and the poor performers, with negative or very low agricultural growth (see table B.2, appendix B).

The better performers did not consistently do well over the past decade, however. Some countries moved from being poor performers in 1990–2000 to being better performers in 2000–04, and some moved in the opposite direction. The change has often been dramatic, which makes aggregate growth rates misleading. For example, agriculture in Angola grew at 13.7 percent a year during 2000–04, but had retreated by 1.4 percent during 1990–2000. The high growth in the later period was because the country was starting from a very low base after a period of conflict.[1] Only about a dozen countries, among them Benin, Burkina Faso, Ghana, Nigeria, and Tanzania, show consistent growth in agriculture of over 3 percent over the period 1990–2004 (table B.3, appendix B).[2]

Agricultural sector growth has been highly erratic across the Region and over time.

Agricultural production in Africa has grown since the 1960s, but that growth is distinctly different from that in other Regions. Great strides in cereal production in South Asia over the 40-year period from 1961 to 2001, for example, were mainly the result of increased yields (figure 2.1 and table B.4, appendix B). African production of both cereals and root crops in the

Increased agricultural production over 1961–2001 was mainly the result of more land under cultivation.

Figure 2.1: Changes in Cereal Production Produced by Changes in Area and Yield, 1961–2001 (1961 = 100)

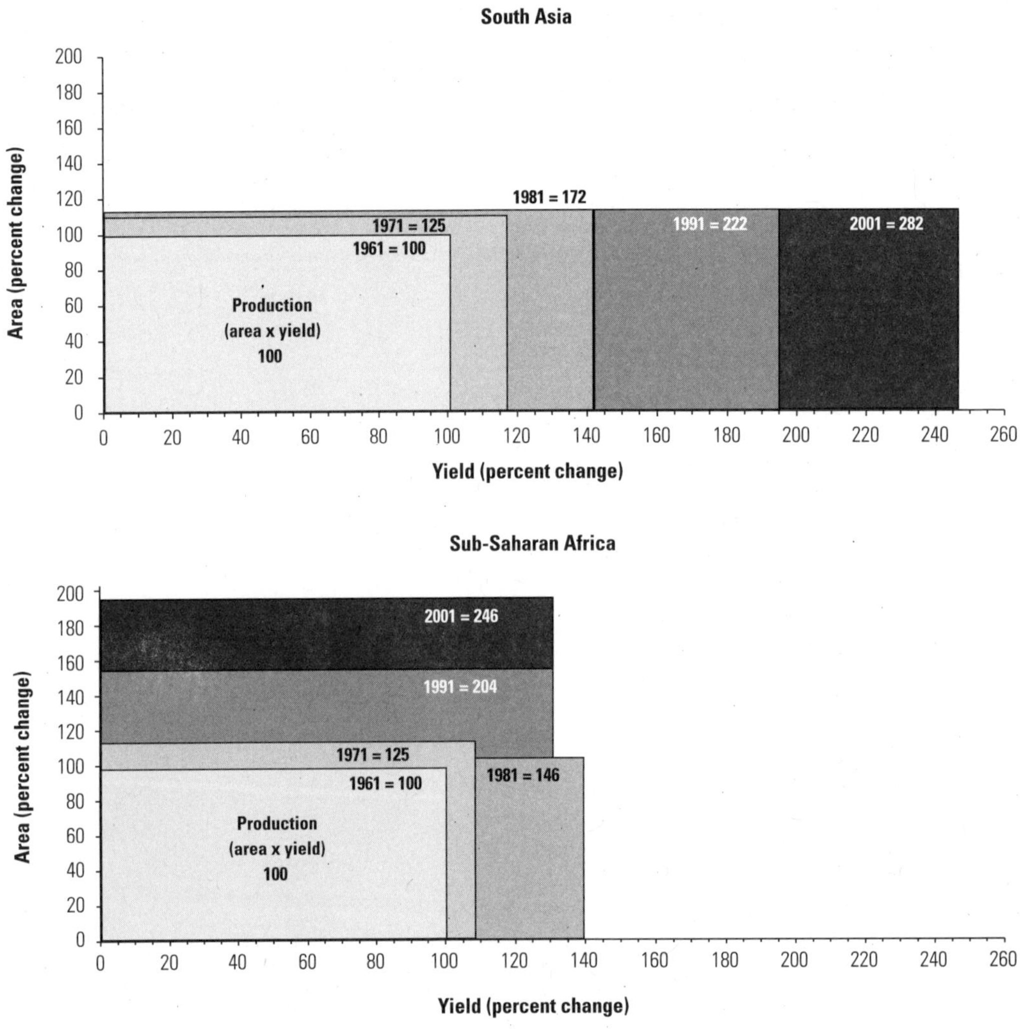

Source: Henao and Baanante 2006.

same period rose mainly because more land was brought under cultivation, while crop yields were largely stagnant (Eilitta 2006). In recent years, however, expansion too has stagnated, indicating that land frontiers may have been reached, at least in some countries.

The rapidly increasing population has also further reduced the arable land per capita. Paradoxically, even with rising population numbers, the high incidence of HIV and AIDS and diseases such as malaria have created shortages of labor for cultivation in several countries (World Bank 2000; Shapouri and Rosen 2001). However, the implications of this capacity issue need to be examined much more systematically (IFPRI 2004b).

Food imports have grown rapidly over the period of fiscal 1991–2006. Food production in the Region as a whole has not kept pace with population growth, and food imports have filled the gap. Meanwhile, Africa's exports, which are primarily agriculture-based, have declined, and in several commodities, including coffee, the Region has lost its share of the world market to competitors. Beginning in 1973, Africa became a net food importer, and this represented the beginning of a chronic food gap for the Region (Eicher 1999).

The Aid Architecture for Agriculture in Africa

Both multilateral (World Bank, IFAD, FAO, AfDB) and bilateral (such as Development Cooperation Directorate–Development Assistance Committee, or DCD-DAC, member countries) donors have provided support for agriculture development in Africa. However, aid to African agriculture from both sources declined between 1981 and 2001 (appendix E). With the increasing focus on the development of Africa, both bilateral and multilateral aid to African agriculture has picked up since 2000. More recently, China has become an important bilateral donor to African agriculture. Average annual aid flows to Africa as a whole were 13 percent higher in 2000–05 than in 1995–2000 (UNCTAD data).

Both bilateral and multilateral donors have been equally important players in terms of aid amounts provided. Organisation for Economic Co-operation and Development (OECD) data show that though bilateral donors as a group have played a comparatively larger role, the Bank (IDA [International Development Association]) was the single largest donor to African agriculture over the period 1990–2005. The largest bilateral donors were the United States and Japan (table E.2, appendix E). Twenty-five percent of Bank-supported projects in the agriculture sector have been cofinanced by other bilateral and multilateral donors.

Foreign private sector flows into Africa are modest in comparison with bilateral and multilateral aid (Hazell and von Braun 2006). Of foreign direct investment (FDI) in the developing world as a whole, less than 1 percent went to Africa in the early 2000s (IFPRI 2002a). Africa's connections with the modern global economy are weak, and private commercial investment in agriculture has been largely limited to export crops and higher-potential zones. Even here, while international commodity markets have continued to expand, Africa's exports have shrunk over time, and today Africa's total volume of exported farm commodities (groundnuts, palm oil, and sugar, among others) is actually smaller than it was 30 years ago (IFPRI 2002a).

As productivity stagnated, food imports increased.

Some nontraditional exports—such as Kenyan flowers, Nigerian shrimp, Malian mangoes, and pineapples and beans in several countries—have fared well. Private investment in agricultural research and development (R&D) has been small; it was only about 2.3 percent of the total spent on R&D in 2000, and much of that was spent in South Africa. A number of international seed companies have invested in maize seed multiplication, and in September 2006 the Rockefeller and Bill and Melinda Gates Foundations together launched a new partnership to help Africa develop its agriculture.

Nongovernmental organizations (NGOs) have also been increasingly involved in African agricultural development, particularly in activities that involve community mobilization and extension support services. Some NGOs have also been participating in research and the development of marketing chains and input supply. However, the effectiveness of NGOs in contributing efficiently to development in these areas has still to be assessed.

The Bank has been an important player in the overall aid environment for agriculture, although both bilateral and multilateral donors have been important.

Donor coordination

A major challenge has been the varied strategies and priorities of the bilateral and multilateral donors that provide support for agricultural development in Africa. The literature suggests that over the years, there has been some improvement in coordination among donors, but more so on procedures than on policies and strategies.[3]

Donor coordination of strategies still has a long way to go.

The country is expected to be in the driver's seat on the strategy for development of a sector. Though progress has varied across countries, there is little systematic evidence to suggest that Bank support for agricultural development is part of a coordinated approach among donors to support country strategies for development of their agriculture sectors. A review of the Bank's Country Assistance Strategies (CASs) carried out for this study found that two-thirds of the documents do not discuss coordination of agriculture interventions by donors. Of those that do, there is little detail on specifics. In other words, while commitment to donor coordination is signaled, the form of the relationship between Bank and other donor interventions is not.

A review of the sample of project appraisal and completion reports also found that while there is some discussion of intent to coordinate particular donor activities at the appraisal stage, there is little follow-through. At the completion stage, the reports provide little or no information on other donor support in the area, or how the Bank effort fits in with the activities of other donors in agriculture. Completion reports for Bank projects rarely, if ever, report on the activities supported in the same project by other cofinanciers.

The World Bank's Strategic Approach

The Bank has no separate agriculture strategy for Africa—its approach has been embedded in the Bank's broader rural development strategy.

The Bank has no separate strategy for the *agriculture* sector, but rather has usually articulated its agriculture goals in the context of broader rural development strategies. There also have been several subsector strategy papers and operational directives, such as those for forestry and water resource management.

In the Rural Development Strategy Papers, the importance given to agriculture has varied over time. A review of the three rural strategy documents (*Rural Development Sector Strategy Paper,* 1975; *Vision to Action,* 1997; and *Reaching the Rural Poor,* 2003) revealed that agriculture had greater prominence as part of rural development in the 1970s than in later years, mainly because in the initial years, the Bank's activities in rural areas were primarily related to agriculture. The Bank's 1975 Rural Development Policy Paper (World Bank 1975, p. 18) noted:

> *The central concept of rural development presented here is of a process through which rural poverty is alleviated by sustained increases in the productivity and incomes of low-income rural workers and households. . . . Most of the low-income groups in the rural areas depend heavily on agriculture for their livelihood. It follows that many of the programs intended to raise rural incomes must center on agricultural development.*

In the mid-1980s, the Bank began to expand its role in human development, and environment and sustainable development became important concepts in the mid-1990s. The next rural strategy, *Vision to Action* (1997), took on a broader rural focus,[4] which persisted in *Reaching the Rural Poor* (2003). This led to increases in Bank rural lending over time, and agriculture became a smaller percentage of the total rural portfolio. The timing of this shift had important implications for donor support for agricultural development in Africa, as discussed in chapter 1.

The Bank has not had a formal agriculture strategy document for the Africa Region, though some technical and discussion papers were produced and were influential in shaping strategic thinking on agriculture in the Region. The 1993 paper *A Strategy to Develop Agriculture in Sub-Saharan Africa and a Focus for the World Bank* (World Bank 1993c) emphasized reform of the enabling environment to enhance private sector interest and restructuring of parastatals and other services where private operation is likely to be more efficient. It also encouraged more regional integration of agricultural markets and put more emphasis on land tenure. Both *Vision to Action* and *Reaching the Rural Poor* included specific development strategies for Africa, and both recognized the importance of increasing agricultural productivity for agricultural development.

More recently, the World Bank's 2005 *Africa Action Plan* (World Bank 2005e) recognized the agriculture sector as a potential driver of growth. The Comprehensive Africa Agriculture Development Programme (CAADP) is at the heart of the New Partnership for Africa's Development (NEPAD) initiative to accelerate growth and eliminate poverty and hunger. The Africa Action Plan, in line with the CAADP, gives priority to making agriculture more productive and sustainable. Among other things, the Action Plan emphasizes increasing public and private investments to expand irrigation by 50 percent over the fiscal 2005 base by the end of fiscal 2008, with the Bank as lead financial partner. NEPAD also advocates Regional integration to overcome the fragmentation of the continent and to reduce Africa's economic marginalization. The Bank's Action Plan recognizes the importance of supporting these initiatives.

From the various rural strategy documents, this review extracted the broader strategic goals the Bank has pursued in African agriculture during fiscal 1991–2006. A wide range of issues is covered, as reflected in table A.1, appendix A. The treatment of issues differs across documents. Moreover, there are inconsistencies among priorities in the different documents. For example, it is not clear why the Africa Action Plan makes irrigation a priority, when two years earlier the Regional strategy in *Reaching the Rural Poor* emphasized that rain-fed agriculture should take priority since "over 95 percent of cultivated land is rain-fed . . . increasing yields on rain-fed lands by just 10 percent would have far greater impact on total agricultural output than doubling area under irrigation" (World Bank 2003d, pp. 101–02).

The recent "Progress in Implementation" report on the Africa Action Plan (DAC 2007) rightly emphasizes the importance of increasing agricultural productivity in Africa, though it is not clear how much importance it is accorded relative to other priorities identified in the Action Plan, given that progress is lagging. The progress report clearly notes that "the AAP [Africa Action Plan] is on track to meet the expected outcomes in all but two (agricultural productivity and gender) of the [shared growth] pillar's nine thematic areas" (DAC 2007, p. 6). The report also implies that it will support both irrigation and rain-fed agriculture, but it is not clear how limited resources will be distributed between the two and how adequate resources will be mobilized to meet the anticipated outcome of an "increase in irrigated land by 2011," which has replaced the 50 percent target noted above. It is not clear how progress toward the "increase in irrigated land" is to be assessed without a target.

With the broader rural focus from the 1980s onward, agriculture received an increasingly small share of lending.

From the comparative analysis of the strategy documents, IEG identified a set of critical constraints to agricultural development in Africa that were defining the Bank's strategic approach. A review of the literature provided further support that these constraints were key to the development of agriculture in Africa. The constraints are as follows:

- Agro-ecological diversity
- Rainfall and droughts
- Soil fertility
- Water
- Seeds
- Credit and rural finance
- Transport infrastructure
- Extension
- Land reform
- Price and marketing reform.

Strategic documents include a number of broad goals for agriculture. From these, it is possible to identify the set of constraints to agriculture in Africa that have defined the Bank's lending and nonlending program in the Region.

The constraints are detailed below and used in the evaluative review of the Bank's performance in chapter 5. In addition to the above constraints, the study covers Bank and borrower capacity issues, including building research capacity, in chapter 4.

Some issues that appear as a priority in the strategy documents are not covered as standalone issues in the thematic assessment in chapter 5. These include issues related to agroforestry, agro-business, livestock, and natural resource management. Gender, the importance of which is acknowledged in the strategy documents, is not covered separately but is treated where appropriate. Finally, decentralization and empowerment of producer organizations are not addressed because they are part of other IEG studies.

Main Constraints to Africa's Agricultural Development

Agro-ecological diversity

Sub-Saharan Africa has a total land area of 2,455 million hectares, 41 percent of which is classified as agricultural land. The Region is characterized by a diverse range of agro-ecological zones spread across countries. A country can include land area that falls under several agro-ecological zones, as in Ethiopia, for example. The arid and semi-arid ecological zone in Africa accounts for 43 percent of the land area; the dry subhumid zone, 13 percent; and the moist subhumid and humid zones, 38 percent (FAO 2001).[5] Based on the natural resource base, dominant livelihood, and the degree of integration between crops and livestock, several production/farming systems with variable potential for agricultural production have been defined for the Region (see table C.2, appendix C).

The Region has a diversity of agro-ecological zones and differentiated production and farming systems.

Rainfall and droughts

One of the biggest challenges faced by the average smallholder in Africa is food insecurity arising from risk of crop loss from variations in rainfall and droughts. Climatic variability is a particular problem in the arid and semi-arid ecological zones. Even in years when precipitation is adequate overall, rain can start late or finish early, with disastrous consequences for agriculture. Rainfall variability in Africa is roughly twice that of temperate regions (World Bank 2004a). Droughts in the Region are also much more frequent than anywhere else in the world.[6] Pests and diseases add to the vulnerability faced by farmers. For example, invasions of desert locusts have occurred repeatedly in the Sahel region and have triggered famines in several West Africa countries.

Africa has a high degree of climatic variability, and droughts are more frequent than in other Regions.

To survive in such a harsh environment, farmers must rely on diversified coping strategies, which influence decisions about the choice of crops planted, inputs used, and non-farm activities taken up.[7] Unlike farmers in South Asia, where irrigation is widespread, most African farmers do not produce a single crop such as rice or wheat in one season. Instead, to ensure at least some produce from their land, farmers normally plant several varieties of crops (typically 10 or more) with different maturity periods, together with trees. Millet, sorghum, maize, cassava, and other root crops are among the most important food crops in the Region.[8] Cereals such as rice and wheat, the mainstay of Asia's Green Revolution, are grown, but are less important. Livestock rearing is also a critical part of this diversified system and is a source of wealth to be drawn on for survival when all else fails.

Soil fertility

Low soil fertility is a major contributor to the low productivity of African production systems (Sanchez and others 1997; Donovan and Casey 1998; Scoones 2001; Mekuria and Waddington 2002; and Sasakawa Africa Association 2004a). Only 6 percent of the land in the Region has high agricultural potential (Tegene and Wiebe 2003 quoted in Ehui and Pender 2005).

Soil fertility is affected by a number of factors. Compared with soils in parts of North America, Europe, and Asia, most African soils are naturally low in nitrogen and deficient in phosphorous, sulfur, magnesium, and zinc (Grant 1981 quoted in Donovan and Casey 1998). In addition, most parts of Africa have shallow topsoil that provides little root room for crop anchorage and extraction of nutrients and water (ECA 2003). Soils are also heavily leached and have high acidity and low organic content (Donovan and Casey 1998). Poor soil fertility was less critical for agricultural development when it was possible to freely extend the land frontier and allow some agricultural land to lie fallow. However, rapidly growing populations and land shortages have reduced the amount of potential fallow land, as well as the length of fallow periods, further reducing soil fertility.

Of course, soil fertility can be improved by the application of organic and inorganic fertilizers and better land management practices, including application of indigenous techniques to increase soil fertility and water retention, such as tie ridges. But that has not happened in Africa. Unlike other continents where soil fertility depletion has been tackled by applying fertilizers, Africa has had tremendous quantities of nitrogen and phosphorus taken out of the soil that have not been returned (IFPRI 2004b). Labor shortages also often deter farmers from investing in indigenous low-input intensification methods, and organic fertilizers are not available in large enough quantities to provide the necessary basic nutrients (Sanders and others 1996). Cattle diseases and shrinking farm size have limited access to organic fertilizers for many farmers, which increases the need for inorganic fertilizers (SIDA 2006).

Most of Africa relies on imported fertilizers purchased at highly variable international prices, and poor infrastructure adds to fertilizer, distribution, and marketing costs, putting it out of reach of most farmers. Fertilizer costs per ton average out to a farmer price of $336 in Nigeria, $321 in Malawi, $333 in Zambia, and $828 in Angola, compared with $227 in the United States (Eilitta 2006). In the era before adjustment lending, many African countries relied on subsidies to get fertilizers to farmers at a reasonable price. With the removal of subsidies, fertilizer prices have soared.

Lack of access to water also makes farmers reluctant to use fertilizers, since their application without water increases the risk of crop failure (Camara and Heinemann 2006).[9] Consequently, the average intensity of fertilizer use throughout Africa remains much lower than in other Regions—roughly 9 kilograms per hectare versus 86 kilograms in Latin America, 104 kilograms in South Asia, and 142 kilograms in Southeast Asia and has been virtually stagnant during the past decade.[10]

Water

The majority of the soils in the continent have poor capacity to hold and release moisture. As a consequence of the variable rainfall and poor soil quality, it has been estimated that about 86 percent of Africa's land area is under moisture stress. Moreover, water conservation and management in rain-fed areas is not practiced adequately.

A major constraint on productivity is low soil fertility—only 6 percent of the land in Africa has high agricultural potential.

Fewer than 5 million hectares of the land in Africa are irrigated—about 4.9 percent of total cultivated area compared with 40 percent in South Asia. More than 3 million hectares of that irrigated land are in just two countries—Madagascar and the Sudan (Wiggins 2000). Agricultural production in most parts of Africa is carried out without irrigation. The area under irrigation is a very small part of the potentially irrigable area in most countries (table K.1, appendix K), which also have limited water storage infrastructure.

Only about 5 percent of the cultivated area is irrigated and 86 percent of the land area is under moisture stress.

A major constraint on expanding irrigation infrastructure is the high investment costs, ranging by one estimate between US$5,000 and US$25,000 per hectare, much higher than in Asia (quoted in IFPRI 2005a), though a recent study by the International Water Management Institute argues that it is possible to design and implement projects in Africa with unit costs comparable to those in Asia.

Further, a large part of the area currently under irrigation is low-performing because of poor maintenance of irrigation schemes, inadequate attention to improving water reliability and control, low use of inputs, and lack of access to market, among other things (Peacock, Ward, and Gambarelli 2007).

Seeds

Sustained use of high-yielding seed varieties was the driving force of the Green Revolution in Asia. In Africa, research has also contributed to development of improved varieties for most of the important food and cash crops over the past 20 years. High-yielding varieties of maize

Organic fertilizers are in short supply and inorganic fertilizers are very costly.

Improved seed varieties have been developed but are not widely used for a variety of reasons.

and new rice varieties (New Rice for Africa, or NERICA) that are also early maturing, pest and disease resistant, and drought tolerant have been heralded as important successes in several areas. However, widespread and sustained use of improved varieties has been constrained by limited availability of inputs and credit, inadequate extension, and the wide variation in required characteristics across multiple agro-ecological zones.

Credit and rural finance

Almost all countries in Africa have a large unmet demand for agricultural credit and rural finance. With inadequate financing in the short term, farmers find it difficult to buy inputs and seeds. In the long term, they are unable to invest in land improvement, better technology,[11] or irrigation development. Improving the provision of and access to financing for agriculture can meet a range of needs and can be critical to the success of agricultural development programs (World Bank 2005c).

Most countries in the Region have high unmet demand for credit and rural finance, so farmers find it difficult to buy inputs and seeds or to invest in longer-term improvements.

Before the era of adjustment lending, governments in several countries ran a variety of input credit programs, which led to huge government deficits because of poor repayment rates (Kelly, Adesina, and Gordon 2003). During the adjustment phase, many of these programs were abandoned. In addition, one result of the adjustment reform agenda was the privatization of parastatals responsible for marketing of crops such as cotton. These parastatals used to meet the credit needs of the farmers for inputs, and their privatization also left a gap that has not been filled. In contrast to conditions in Asia, there are few specialized moneylenders in most of Africa (Collier and Gunning 1997). Moreover, because of the existence of several constraints (box 2.1), new sources of credit for smallholders have been slow to develop.

However, the difficulty of providing farmers with access to credit does not mean that there can be no viable and sustainable institutional modalities to provide credit to smallholders in Africa's difficult environment. Recent research from the Consultative Group to Assist the Poorest (CGAP 2005) demonstrated that there may be successful microfinance providers for agriculture, though this issue needs further analysis. The

Box 2.1: Constraints to Development of Access to Credit and Rural Finance in Africa

Supply Side:

1. High, interrelated covariant risks created by variable rainfall and lack of irrigation, pests and diseases, price fluctuations, and constrained smallholder access to inputs, advice, and markets
2. Small size of farms and of individual transactions
3. Dispersed demand for financial services because of low population densities
4. High transaction cost for service providers because of remoteness of clients and heterogeneity among communities and farms
5. Seasonality of agricultural production leading to lag between investment needs and expected revenues
6. Lack of usable collateral because of ill-defined property and land-use rights, high cost or lengthy registration procedures, and social constraints to foreclosure
7. Underdeveloped communication and transportation infrastructure
8. Weather and price risk (both a supply- and demand-side constraint).

Demand Side:

9. Low affordability of market interest rates for farmers
10. Insufficient cash-flow planning
11. Repayment schedules are often difficult because they are not adapted to seasonality of the crop cycle
12. Weather and price risk.

Sources: World Bank 2005a, 2005c, 2005d; study research.

CGAP research notes some of the special features of these providers that can help overcome the challenges noted in box 2.1: delinking repayments to loan use, character-based lending techniques combined with technical criteria in selecting borrowers, providing saving mechanisms, diversifying portfolio risk, adjusting loan terms, and conditions to accommodate cyclical cash flows, among others.

Transport infrastructure

Perhaps the most critical of the remaining barriers to market access in Africa is inadequate transport infrastructure. Unlike Asia and Latin America, Africa inherited a highly dispersed and unevenly distributed infrastructure from its colonial past (IFPRI 2005a). In most African countries, less than one-third of domestically produced food enters commercial marketing channels beyond the local area (Sasakawa Africa Association 2004a). In one indication of the severity of rural farmer isolation, Hine and Rutter (2000) estimate that for 51 percent of villages in Ghana and 60 percent of those in Malawi, the walking distance to the nearest pickup point for motorized transport services was more than 2 kilometers; it was over 10 kilometers for 10 percent of Ghanaian villages and 19 percent of Malawian villages.

IEG's recent Transport Sector review (IEG 2007o) found that transport costs account for 11.5 percent of the total value of imports in Africa, compared with 7.2 percent in Asia and 6.7 percent in North America. On the export side, for many countries in Africa, at least 20 percent of the export costs are directly attributable to transport. For landlocked countries such as Malawi, the figure can be as high as 55 percent. This very seriously weakens the terms of trade for such countries. On the basis of their work on growth, distribution, and poverty in Africa, Christiaensen and others (2002) found that whether a household has access to infrastructure and urban markets was immensely important in governing the growth in household income.

Extension

Inadequate farmer access to improved technologies and land management practices has proved to be a major constraint in Africa, and the literature has identified a number of cases, including cassava, sweet potato, millet, and rice, for which high-yielding varieties are underutilized or farmers are operating within the production frontier (Evenson and Gollin 2003; Christiaensen and Demery 2007). Despite the tremendous need, most of the extension approaches that have been tried have met with limited success. Moreover, a major part of food production is undertaken by women farmers, and in the past most extension systems have not tailored their extension approaches to women's specific needs.

There may be viable and sustainable institutional modalities for providing credit under the difficult environment in Africa.

Most extension approaches that have been tried in Africa have had only limited success.

Price and marketing reform

One of the main reasons that price and market reforms are needed is that the incentives for agricultural production are weak. Both price (output and input prices) and nonprice factors (access to markets, credit, among others) determine farmers' incentives to produce. Primarily because of limited access to markets, because of the transport constraint, the majority of smallholders produce largely for self-consumption. In areas with reasonable market access, cash crops also become attractive, though the possible returns on both food and cash crops determine the extent to which a farmer produces one over the other. However, several domestic market distortions and subsidies in OECD countries have prevented farmers from getting good returns on crops they market. Marketing and other reforms were meant to improve the incentives for farmers by reducing domestic market distortions and by encouraging private traders to substitute for inefficient state trading companies (as discussed further in chapter 5).

Transport constraints limit market access, and market distortions reduce returns on cash crops.

Land

Formally codified property rights regimes are still quite rare in Africa, and most land falls under customary law (van den Brink and others 2005), although the situation varies considerably by

Formal property rights are rare in the Region, and women typically have to negotiate through male relatives.

country. Much of the land under customary law is considered state-owned, and as land becomes a scarce resource with increasing population, outsiders may be able to appropriate the land through misuse of land titling laws. Land is also a key patronage resource to reward political favors, and security of land tenure can be affected by political decisions (IFPRI 2004b). This can make agricultural development a very sensitive political issue.

Several governments, including those of Ghana and Uganda, have sought to address this constraint through broad recognition of customary rights, but progress has been slow. Although women typically conduct the majority of the farm work in Africa, they rarely have full rights to land, but must negotiate as secondary claimants through a male relative (Toulmin 2006).

Nature of agricultural development in Sub-Saharan Africa

Agricultural development is multifaceted. It requires coordinated interventions across a range of activities, both within the sector and in other supportive sectors, to deal with the constraints noted above. More than any other sector, the development of agriculture requires the activities of various subsectors or other relevant sectors to contribute effectively *at the same time, or at least in some optimal sequence.* For example, it is difficult for farmers to buy inputs unless there are functioning credit institutions to meet their credit needs. Markets cannot be accessed if the roads are poor, and farmers cannot know about improved technologies or participate in adaptation if good extension is not in place. Soil fertility improvement requires not only access to improved technology, but also improved inputs, including water.

Agricultural development in Africa requires coordinated interventions across a range of activities.

With the right inputs, infrastructure, incentives, and technologies, rapid growth in agriculture incomes is possible in Africa.

Several of the challenges in Africa today were not major factors in Asia when that Region was developing its agriculture, because countries such as India already had a critical minimum of infrastructure, irrigation, and industrial capacity to produce fertilizers, among other things, and, with the improved seeds that came with the Green Revolution and extension, agriculture took off.

Hence, support for agricultural development in Africa needs to appreciate the challenges that are specific to Africa. While the broader rural focus of the Bank from the mid-1980s onward was justified, an unintended result was that it led to less focused attention on the need for various activities that are critical to agricultural development in rural space to come together at the same time, or at least appear in some optimal sequence.

Development of agriculture in Africa is complicated even further by the risk factor in agriculture. For example, increasing the availability of hybrid seeds will not ensure that the seeds are actually used unless farmers are convinced that the increased output would not come at a higher risk. Exposure to droughts and weather-related uncertainties affect a farmer's incentives to adopt high-risk technologies, and they may often forgo technologies that would require them to use fertilizers that would yield higher outputs, but present higher risk (Dercon and Christiaensen 2005). While this would be an important consideration in a farmer's decision-making process in other Regions as well, the frequent droughts and low irrigation in Africa make the environment in most areas in the Region riskier.

The Millennium Development Project's Hunger Task Force (UNDP 2005) concluded in 2005 that the world could meet the MDG of halving hunger by 2015. Development of smallholder agriculture in Africa is critical to that goal. The literature shows that with the adoption of improved technologies and modern techniques, access to agricultural inputs, and investment in infrastructure, rapid growth in agricultural incomes is achievable in Africa (Howard and others 1999; Palmer 2004). Smallholder agriculture, which is the predominant source of livelihoods in Africa, has proven to be at least as efficient as larger farms when farmers have received similar support services and inputs (seed, fertilizer, and credit) (IFPRI 2002b). Sustaining success, however, has often been problematic (Wiggins 2005). The diverse African situation also implies that no single solution will radically improve African agriculture and a comprehensive set of strategies will be needed (InterAcademy Council 2004). Most success stories involve measures that address the vulnerability, volatility, and risk in the sector (Commission for Africa 2005).

Chapter 3

Evaluation Highlights

- Agricultural analytical work has fallen short of its potential to inform policy dialogue and lending.
- Policy advice associated with Bank-financed adjustments has had far-reaching implications for agricultural development in Africa.
- The Bank's limited lending has been fragmented and did not properly recognize the multifaceted and interconnected nature of agricultural activities.

Woman watering a field in Ghana. Photo by Curt Carnemark, courtesy of World Bank Photo Library.

Bank Support for Agriculture and Portfolio Performance

The Bank's activities in support of agricultural development in Sub-Saharan Africa fall into three broad categories: analytical work, policy advice, and lending. Of the three, the analytical work is perhaps the most critical for the diagnosis of issues and the suggestion of possible solutions. It is meant to inform both policy advice and lending.

Analytical Work

Over the review period, the Bank has produced an array of analytical products relevant to agriculture in Africa. Some of this work has focused broadly on the Region, some on particular country issues. Some has addressed the whole agriculture sector, some has concentrated on subsectors, such as extension. Still others have looked at specific commodities, such as cotton, coffee, tobacco, and cashews. Much of the analytical work has been produced by the Bank's Africa Region and Agricultural and Rural Development Department (ARD), but the Bank's Research Department has also done several studies. Since the Trade Department was created in 2002, there has been a considerable increase in the number of trade-related analytical studies relevant for agriculture.

Quality and quantity of analytic work

Despite the apparent variety of analysis done on agriculture in Africa, it is not of sufficient quantity or quality. Reviews by ARD and the Bank's Quality Assurance Group (QAG) indicate that analytical work for agriculture in general has been of insufficient quantity. However, in keeping with the emphasis on increased analytical work in the Bank's 2003 Rural Strategy (World Bank 2003d) and recent increased interest in agricultural development in the Region, the quantity of analytical work has increased in recent years, though it has been spread unevenly across countries.[1] That said, regional and global partnerships could augment resources for analytical work, particularly in small countries.

The quality of the available analysis is variable, though it has been improving, as noted in QAG annual reviews since the late 1990s, when Bank management recognized this issue as a concern. The agriculture portion of multisector analytical work, such as Public Expenditure Reviews (PERs), has also been weak.[2] One of the strongest areas of analysis at present appears to be in trade. Much of the

The quantity of agriculture-related analytical work in Africa has been increasing recently, but it is unevenly distributed.

The quality of agriculture-related analytical work has been variable but is improving.

work in this area has been produced to back the Bank's efforts in lobbying for a genuinely pro-development Doha Round and for eliminating OECD agricultural subsidies.

Analytical work and policy dialogue and lending

Regardless of the quantity or quality of analytic work in agriculture, however, that work is of limited use if it does not adequately influence Bank lending or policy dialogue. While the available analytical work emphasizes the importance of agriculture to development in Africa, it does not appear to have adequately informed the lending and policy dialogue relevant to agricultural development in the Region. Global reviews (which include Africa) of analytical work done by QAG have also found this shortcoming.[3] A recent Quality Assurance Review of agriculture-related analytical work noted the rather low level of importance assigned to ARD analytical work in country programs.[4] The portfolio review for this study also found that only about one-third of the Project Appraisal Documents noted that the design had been informed by a piece of analytical work. This finding was also supported by the staff survey done for this study. More than 55 percent of the survey respondents agree that sufficient and rigorous analytical work generally does not inform the design and implementation of agriculture projects in Africa.

Agriculture-related analytical work appears to have had limited influence on lending and policy dialogue.

Even the Bank's most recent trade-related analytical work has not had much influence on lending or country dialogue. A recent IEG study (IEG 2006a) found that outside observers associated with World Trade Organization negotiations thought that while the Bank was an important player in generating research relating to the negotiations, the Bank's research did not find practical application at the country level.

Analytical work has not helped to prioritize lending based on changing country-specific needs.

Analytical work has also not been able to help prioritize or sequence lending according to changing country-specific needs, as acknowledged in assessments undertaken by QAG.[5] Such findings also emerge from IEG work. For example, reporting on the weak quality of the agriculture strategy note, the Rwanda Country Assistance Evaluation (IEG 2004a, p. 19) notes:

> *The Rwanda program was not unique in this respect. An internal assessment in the late 1990s of Bank-wide economic and sector work provides a partial explanation of why analytical work may have received relatively fewer resources than lending activities. It noted that economic and sector work was weakest in Africa and in the Latin America and Caribbean Regions and it offers a conclusion which applied to the entire Bank. "Finally it is often unclear what the priority of [economic and sector work] is within the Bank. Too often task teams feel that their ESW responsibilities are secondary to those of preparing lending operations. As a result, ESW timetables often are the first to be dropped or postponed during crunch periods. With staff typically over programmed, ESW tends to get lower priority and quality can suffer because of this."*

This also partly explains why few African countries have consistently had analytical work produced over time.

There are four reasons that analytical work does not appear to have adequately informed Bank lending and policy advice.

First, analytical work has been of limited quantity and not easily available, even within the institution, principally because of inadequacies in the Bank's databases. QAG reviews of analytical work confirm this finding.[6] The Bank's database does not even have a systematic record of all agricultural and rural analytical work produced in Africa. In undertaking the Mali country review for this study, for example, it was very difficult to locate agriculture-specific analytical work, and staff in the Region confirmed that several pieces had not been entered in the Bank's database. Further, there are no records in the Bank's databases for informal analytical work produced as an input to the preparation of a project. In a knowledge-based institution such as the World Bank, it is surprising that the record of analytical work is so poor.[7]

Second, interviews with Bank staff reveal that the incorporation of findings from analytical work in lending and policy dialogue is not functioning well. While Decision Meetings are supposed to be the forum to ensure that analytical findings are incorporated in project design, requiring at least some peer reviewers to explicitly comment on the extent to which a project proposal responds to available internal and external analytical findings might help to strengthen the linkage. Another option may be to institute a more formal record, similar to the IEG/Bank management tabulation of the Management Action Record or some other formalization. The incorporation of findings from analytical work currently depends too much on individual staff or peer reviewer interests and shifting country or thematic institutional memory.

Third, the sectoral organization of the Bank has impeded interaction among staff across sectors. As a result, good quality analytical work produced in other relevant sectors, such as trade and transport, is also not adequately considered in informing agricultural lending. QAG reports on analytical work confirm this finding.[8] Further, internal reviews note that the Bank rarely builds on analytical work produced outside the institution.

Fourth, the technical quality of analytical work in agriculture appears to have suffered from a decline in technical skills within the institution (discussed further in chapter 4). Bank staff have tried to compensate for this skill shortage by hiring outside experts and using cooperative arrangements with organizations such as the Food and Agriculture Organization, but coordination and timely, quality input have been issues.

Policy Advice

Over the past half-century, developing countries have looked to the World Bank not only for financial support but also for policy advice to promote economic and social development. Assessing the policy advice each country has received over the period 1991–2006 is difficult because it is not written down in any document and often is part of the Bank-client dialogue that accompanies the project preparation process. That process itself is often poorly documented. As will be seen in chapter 5, however, some of the Bank's advice, such as that associated with structural adjustment reform, has had far-reaching implications for agricultural development in African countries. But results have fallen short of expectations.

Limited availability has hampered the influence of analytical work, and the procedure to ensure that it informs lending and policy dialogue is not functioning well.

More recently, as a part of NEPAD, the Bank has provided advisory services for trade and policy harmonization and to help to strengthen the capacity of African Regional and subregional economic communities (Kritzinger-van Niekerk and Houdart 2005).

The Bank's organization has inhibited interaction across sectors.

Lending

Overall amounts and trends

During fiscal 1991–2006, the Bank supported 262 projects with agriculture components in Africa. Several of them have been relatively small parts of wider Bank-supported rural activities. Hence, though the total amount invested in projects with agricultural components over the period has been $14.31 billion (about 28 percent of total lending of $50.49 billion to the Region), the lending for agriculture itself has only been about $4.5 billion, 32 percent of $14.31 (table D.1, appendix D).

Of the total agricultural lending of $4.5 billion to Africa, only $2.8 billion (8 percent of the total Bank investment lending to Africa; see table D.1, appendix D) has been investment lending and $1.72 billion has been structural adjustment or development policy lending (DPL).[9] Of the investment lending, $247.2 million has been for emergency recovery. As a result, the amount of Bank funds truly "invested" for development of the African agriculture sector amounts to

Over 1991–2006 the Bank supported 262 projects with agricultural components in Africa.

On average, Bank investment lending in agriculture has only been $67.6 million per country over the last 15 years.

an average of $67.6 million per country of the countries that have had any agriculture investment from the Bank over the 15-year period. This is only a little more than the size of an average loan for an agricultural intervention in Africa over the period 1991–2006 ($55.5 million). Furthermore, that limited lending has been scattered over numerous activities—and thus has been scarcely enough to have sustained impact.

Analysts argue that Bank support for agricultural development needs to be seen in perspective. First, Bank support is often a small part of a larger multidonor effort to develop the agriculture sector. While this may be true, it is important to see this in the context of the overall decline in lending to agriculture over most of the study period from the entire donor community and the weaknesses in donor coordination, as demonstrated in chapter 2. Second, the Bank has contributed to global and regional programs in Africa that supplement resources for agricultural development. But there are no assessments of how these programs supplement country-level interventions. This dimension will be addressed in the Bank-wide IEG study of agriculture scheduled for completion in fiscal 2009.

Figure 3.1: Sectoral Distribution of Investment Lending in Africa, Fiscal 1991–2006

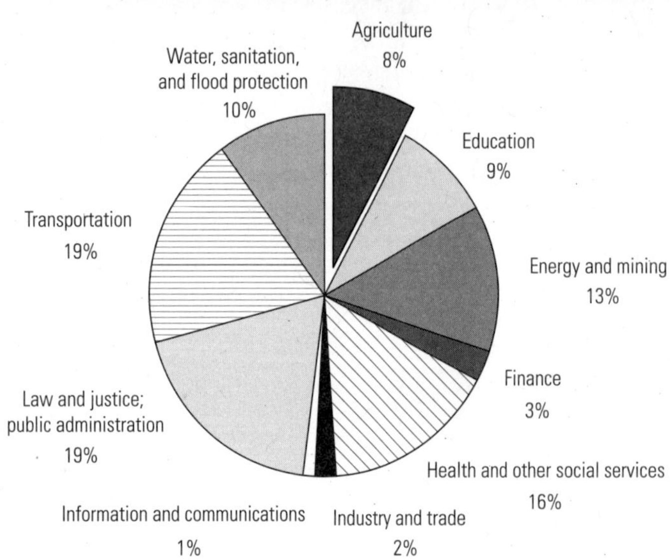

Source: World Bank data.

The Bank's agriculture lending (investment and DPLs) to Africa declined from $419 million in fiscal 1991 to a low of $123 million in 2000 (see table D.3, appendix D for details of the trend in lending). This decline was part of a pervasive trend among donors. In absolute terms, as noted in chapter 2, assistance provided to African agriculture from both bilateral and multilateral donors declined steadily over the 1990s (table E.1, appendix E). Several reasons are given in the literature for this decline, including the high failure rate of many agriculture projects, urban bias, neglect of agriculture by governments, political instability, and a shift in donor priorities toward rural development more widely, among others (IFPRI 1993; OECD 2001; World Bank 2002a; DFID 2004).

Regardless of the reason for the shift, it has meant not only that resources flowing to the sector were inadequate, but also that this downward trend became self-reinforcing. As the decline in lending continued, so too did the decline in recognition within countries that agriculture was central to development in Africa.

> *The success of the Green Revolution [in Asia] also required political support and a favorable macroeconomic policy environment. Foreign aid was helpful in this regard. In the sixties, the governments of most developing countries were largely urban oriented. Agriculture was seen as a holding ground, while the "real investment" in development was thought to take place in the urban, large-scale industrial sector. Foreign aid drew attention to the critical importance of production agriculture in improving the welfare of society. Foreign aid also strengthened the hands of national leaders, who recognized the critical importance of agriculture and of solving the food bottleneck in Asia.* (Mellor 1998, p. 58.)

Bank lending for agriculture (investment and DPL) in Africa picked up beginning in fiscal 2001 and increased sharply in fiscal 2006 to $685 million, up from $295 million in 2005. Presumably this was partly because of the reinvigoration of the Bank's rural programs, as outlined in a new agriculture and rural development strategy in 2003 (World Bank 2003d). It was probably also partly the result of the realization in the international community that Africa was lagging behind and that the agriculture sector is critical to promoting growth and poverty alleviation in the Region.

Major subsectors and country direction

Bank databases do not provide a comprehensive picture of the various activities in the agriculture sector that have received its support. Subsector coding that is expected to provide information on these activities is presented in box 3.1. The Bank's database has eight agriculture subsectors, but these are insufficient to determine the level of support for some critical activities that constrain agricultural development—credit, seeds, tenure, research, and extension, among others. An examination of the existing categorization shows that a "general" category covers about 29 percent of overall dedicated lending to agriculture in Africa during fiscal 1991–2006 (figure D.4, appendix D).

Lending declines through the 1990s led to a decline in recognition of agriculture's importance to Africa.

Based on the categorization in the Bank's database, the second-largest amount of lending over the review period has been for agricultural research and extension (together accounting for 23 percent), followed by marketing and trade (14 percent). Irrigation and drainage together received only 7 percent of total agriculture lending in Africa, although it has been the largest subsector within the agriculture sector Bank-wide (World Bank 2005e). A recent IEG study of

Bank databases do not give an accurate picture of the activities supported by the Bank—the "general" category accounts for the largest share.

Box 3.1: Bank's Coding System and Inadequate Reflection of Important Agricultural Activities

The Bank's categorization system allows task teams to designate up to five subsector codes per project. If project activities cover more than five subsectors, they are expected to use the *general* category. Hence, though the *general* category is a convenient way to manage the data, information on the details of a large share of the lending for agriculture as well as for other sectors is lost. ARD has repeatedly pointed out these problems in the coding system.

The coding system restricts the information available about how much support the Bank is providing to activities that seek to relieve the critical constraints on agriculture in Africa.

Areas Critical to Development of Agriculture in Africa

Access to markets[a]
Irrigation
Drainage
Research
Extension
Credit[b]
Seeds
Incentives for agricultural development[c]
Land tenure

Bank's database subsumes these in different subsector and sector codes.[d] It is difficult to tell how much lending is actually going to these activities.

Bank Subsector Codes for Agriculture

Agriculture extension and research (AB)
Crops (AH)
Irrigation and drainage (AI)
Animal production (AJ)
Forestry (AT)
General agriculture/fishing/forestry (AZ)
Agriculture marketing and trade (YA)
Agro-industry (YB)

a. Roads, which provide access to markets, are coded outside agriculture.
b. Agriculture credit is coded under AZ (above) or under micro- and small and medium-size enterprise finance (FE).
c. Restructuring of Ministry of Agriculture is coded under central government administration (BC).
d. For example, irrigation and draininge are clustered, as are research and extension. Actual amounts for individual activities cannot be distinguished.

water management in agriculture (IEG 2006g) showed that only 3 percent of total Bank commitments to irrigation and drainage between 1994 and 2004 went to Africa.

The largest share of agricultural lending to Africa during fiscal 1991–2006 went to Tanzania (about 10 percent), followed by Côte-d'Ivoire and Uganda. Some African countries (Botswana, Cape Verde, Comoros) have had no agricultural lending over the period. For several others (such as Angola, Democratic Republic of Congo, Guinea Bissau, Lesotho, Mozambique, Republic of Congo, and Sierra Leone), the actual amount of agricultural lending has been very small.

Not 1 country among the top 10 has received consistent and simultaneous support for all critical subsectors.

Not 1 country among the top 10 has received *consistent* and *simultaneous* support across all critical subsectors identified earlier in this report (table D.4, appendix D). The Country Assistance Strategy (CAS) review done for this study also found that discussion of agricultural issues was rarely accompanied by a holistic assessment of the agriculture sector or an explicit indication of how agricultural priorities would be linked to budgets. Only 1 of the 31 CASs reviewed (Ethiopia 1995) comes close to recognizing the integrated nature of relevant agricultural activities to promote agricultural development.

Nearly 83 percent of the respondents to the IEG staff survey agreed that Africa country directors do not sufficiently consider the complex and multisectoral nature of agriculture activities in allocating IDA funds among sectors. That such a multifaceted and cohesive approach toward agriculture is lacking is one factor. Another is that the Bank's data systems do not provide an accurate picture of how much has gone into various critical activities. This limits the extent to which these activities can be meaningfully coordinated.

The outcome rating of agriculture investment projects has been below average, but has improved since 2000.

Overall Performance of Agriculture Projects

As previously noted, in many projects the amount invested in agriculture has been a relatively small part of wider rural activities. To assess the performance of agriculture investments, IEG looked only at closed projects in the Africa portfolio in which the agriculture investment was 50 percent or more of the lending amount. IEG data were used to examine how those projects did in comparison with (a) Africa projects without agriculture components approved in the same period and (b) projects in which the agriculture investment was 50 percent or more from other Regions and that were approved in the same period (figure 3.2).

The review found that over fiscal 1991–2006, about 60 percent of the closed agriculture investment projects in Africa were rated satisfactory on outcome. This rating was below the satisfactory outcome rating of 65 percent for the non-agriculture component projects in the Region. It was also below the 73 percent satisfactory rating for agriculture investments in other Regions (tables D.5, D.6, and D.7, appendix D).

The data show some improvement in outcome ratings since 2000, though the number of closed agriculture investment operations (with an agricultural component greater than 50 percent) is too small to draw a strong conclusion (table D.8, appendix D).

The performance of the Africa portfolio for both agriculture and non-agriculture is worse than in other Regions, although that is hardly surprising, since the quality of the Africa portfolio has lagged behind other Regions for years (World Bank 2004a). But it also suggests that there is more than just the nature of agriculture projects that makes it difficult to achieve satisfactory outcomes in the Region. The literature review, the findings of the country-level agriculture sector reviews, as well as past IEG reviews indicate that political economy, instability, and weak institutional capacity have negatively influenced the outcome of projects in the Region (see chapter 4).

Given the wide variation in agricultural conditions across countries, this review also compared the performance of Bank projects in

Figure 3.2: Outcome and Sustainability Ratings

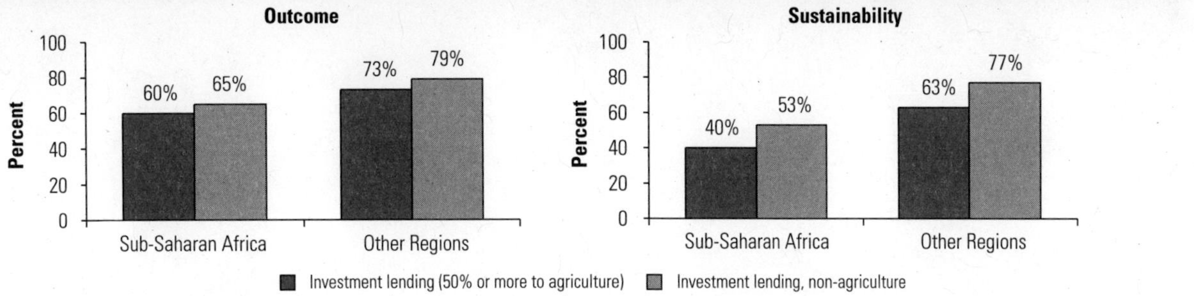

Source: IEG data.
Note: These ratings are for the universe of agriculture and non-agriculture projects, and hence significance tests were not done. But some analysts note that even when presenting results for the universe, it may be informative to apply a statistical significance test to know whether Bank performance or external factors have meaningfully changed over time. The argument made is that satisfactory or unsatisfactory performance of a particular project is partly the product of random or unpredictable factors, and the statistical test would tells us whether the change between years is more than we would expect from random variation, if the underlying chance of each project's satisfactory performance was the same as the mean rate for the year.

countries with more favorable agricultural conditions against those where conditions are less favorable. Surprisingly, as figure 3.3 shows, Bank projects in countries with less favorable agricultural conditions have done better than those in countries with more favorable conditions, although further analysis, possibly with field work, should be done on this issue in the context of the larger IEG agriculture study, because the number of closed projects in countries with less favorable conditions is small. However, the difference in ratings suggests that it is more than factor endowments that are a challenge for agricultural development in Africa.

Figure 3.3: Performance of Bank Projects with Greater than 50 Percent Agricultural Component

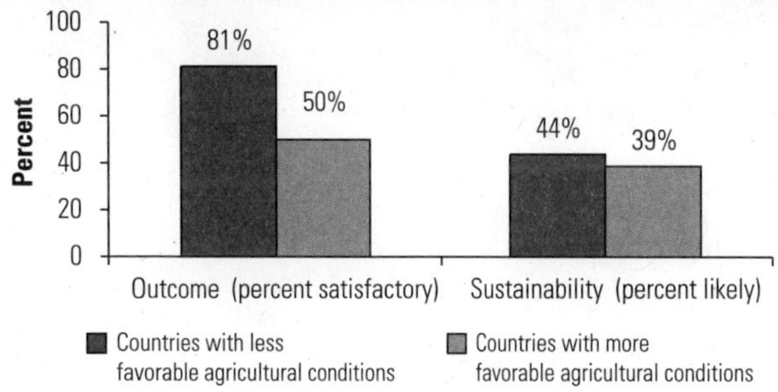

Source: IEG data.

QAG's *2006 Annual Review of Portfolio Performance* also found that the low satisfactory outcome ratings in the Africa Region reflect both *country* factors outside of the Bank's control and *Bank* factors, including a high percentage of fragile states with difficult conditions outside the Bank's control and lower quality-at-entry and supervision ratings. An ARD discussion paper on agriculture and pro-poor growth notes that "while achieving agriculturally led growth faces several key constraints, many of these constraints (such as poor infrastructure and underdeveloped or dysfunctional markets) are also faced by the economy as a whole" (World Bank 2005k, p. 45).

Chapter 4

Evaluation Highlights

- The Bank's internal institutional environment has not been supportive of agricultural development.
- While political commitment in client countries appears stronger now than in the past, overall in-country capacity to support agricultural development is weak and budgetary resources to support agricultural development activities are scarce.
- Research capacity exists, but sustaining and strengthening activities is a challenge.

Boy hand-irrigating field in Burkina Faso. Photo by Curt Carnemark, courtesy of World Bank Photo Library.

Key Factors of Performance

Internal (Bank) Factors

Four factors related to the internal organization of the Bank have influenced its ability to support the development of agriculture in Africa. These are: relations between country and sector units, relations between and within sector units, the technical capacity of the staff in the institution, and the system for monitoring and evaluation of Bank activities.

Relations between country and sector units

The management structure of the Bank distributes accountability and responsibility for the design and implementation of operations between *country* and *sector* staff. Under this "matrix management," the sector units deliver lending and analytical work, but the country management units, led by the country directors, make the decisions about the allocation of resources among competing sector units in their countries.

Task managers of agriculture projects in the Africa Region interviewed as a part of this study noted that there were no well-defined procedures to ensure synergy between the work of country and sector units, and as a result, the agriculture sector was adversely affected. While it can be argued that the CAS process is intended to ensure synergy, the link between the preparation of the CAS and agriculture sector lending and nonlending activities was found to be weak by this review. As noted by a recent IEG assessment of three agriculture projects in Tanzania (IEG 2007k), the strategies for the country and the list of projects financed by the Bank gave the impression of having been developed independently, and then forced together afterwards. A QAG review seems to confirm this finding when it notes that sector studies are frequently undertaken to justify operations in advanced stages of preparation, instead of preceding such preparation efforts (QAG 2004).

Country-level reviews carried out as a part of this study have also noted this problem: the Kenya review found that in several of the CASs for that country, except for the most recent one, the logical connection between the strategy and the lending program in agriculture was not well articulated. For example, in the 1998 CAS, while the El Niño Emergency Project, the roads project, and another agricultural sector investment project were not necessarily incompatible with the country strategy and the CAS objectives, it was not evident that these three choices had been subjected to any rigorous

The agriculture sector seems to have been adversely affected by the matrix management system.

screening process for prioritization. There was no evidence of dropped options. There was a similar finding on Cameroon.

CASs rarely included a holistic assessment of the agriculture sector or linked sector priorities to budget.

This review found further evidence of lack of synergy between the country and sector units in its CAS analysis. As already seen in chapter 3, the CAS review done for this study found the 1995 Ethiopia CAS to be the closest to best practice regarding recognition of the interrelated nature of agricultural development. That CAS recognized that the National Fertilizer Project (1995) and the Seed System Development Project (1995) were designed to improve agricultural productivity and food security through extended use of improved seeds and fertilizer, which were recognized as the two most critical inputs in enhancing yields. Yet there was a complete lack of coordination between the two projects. It seems the CAS logic did not influence the preparation and implementation of those projects. It is thus not surprising that the Africa-specific data on which the fiscal 2003 and 2004 ARD retrospectives of the CASs were based found that the majority of the CASs (57 and 63 percent, respectively) were unsatisfactory in the size and composition of their rural lending and nonlending programs (extracted from communication with ARD, December 12, 2006).

Internal constraints encourage the design of complex projects covering many activities.

The resources allocated to a sector in a country program depend on two factors: the country unit's conviction that a particular sector is worth supporting and demand for investment in the sector from the country. The envelope of available IDA resources is also a constraint, because it defines the upper limit of resources that can be distributed among sectors. Usually no more than one project is supported in a sector in a particular year. This sometimes results in complex project designs, because it creates an incentive for staff to cover as many activities as possible in a given project.

Bank client demand for agriculture lending has not been strong.

An internal review of the quality of supervision for the Lesotho Agricultural Policy and Capacity Building Project notes the tendency to do such complex projects in small countries, because each project may be the only opportunity to work in a sector for years. A similar internal review of the Mauritania Financial and Private Sector Capacity Project (fiscal 1995), which had an agricultural component, also found that the project was trying to do too much in a country with weak administration. The review expressed concern that the project was trying to tackle judicial reform, the mining code, fisheries resource research, the chamber of commerce, and banking supervision—all in one operation.

A recent ARD report (World Bank 2005g) on interviews with country directors, the majority of whom were in Africa, found that their interest in supporting agriculture projects was not very high, since such projects were more time-consuming, riskier, and more expensive to design and implement. The projects were also likely to be more contentious than those in other sectors, especially when they involved forestry or irrigation infrastructure.

Another recent ARD document (World Bank 2005i) acknowledges the complex nature of agriculture projects and their high preparation costs. Data on project preparation costs from the Bank's databases confirms that agriculture interventions in Africa are more expensive than projects in other sectors. Agriculture projects were also found to be riskier (see appendix M). Nearly 63 percent of respondents to the IEG staff survey agreed that supervision and project preparation costs to the Bank for agriculture projects are significantly higher than for projects in other sectors in the Africa Region. Some country directors also found that the rural corporate strategy (*Reaching the Rural Poor,* World Bank 2003d) missed the opportunity to be truly operational (World Bank 2005g). The perception of the country directors and staff was reinforced by the poor performance of agriculture projects in the 1990s and appears to have contributed to the Bank moving away from support for agriculture.

Until very recently, Bank client demand for agricultural lending has not been strong. In a 2003 quarterly report to senior management, the director of ARD noted that many country directors have stated that the decline in purely agricultural investment lending reflects the demand of borrowers for other kinds of support from the Bank, notably adjustment support, which has increased substantially in the Region, as well as the change in agriculture projects to embrace a more community-driven focus. Where there is support for large volumes of investment lending, it is often multisectoral. In the past two years there has been renewed interest in gaining Bank support for agricultural development among countries in Africa, and this is reflected in the consequent increase in lending (see chapter 3).

Relations between and within sector units

IEG's recent evaluation of community-based and community-driven development approaches (IEG 2005a) has drawn attention to ways that the sectoral organization of the Bank handicaps coordination across teams working in different sectors. IEG's *2006 Annual Review of Development Effectiveness* (IEG 2006i) also notes that the Bank's matrix management structure does not encourage staff to work across sectoral boundaries or to address cross-sectoral issues. Agriculture is more susceptible to this problem than any other sector by virtue of its interconnected and multifaceted nature. As already seen, outcomes in the sector are dependent not only on various agriculture-related activities—such as extension, credit, and seeds—coming together, but also on activities of sectors such as transport contributing effectively to agricultural development.

One example of this involves the way agriculture interacts with the transport sector. Respondents to an open-ended question in the IEG staff survey identified lack of rural infrastructure as a fundamental constraint to the development of agriculture. Among the reasons noted for this neglect was the expectation of agriculture staff that rural roads would be covered by staff in the transport sector. However, since there is little coordination across sectors, not much is done to strategically develop rural roads in Bank transport projects to ensure that a Bank agricultural intervention attempting to increase agriculture productivity in a particular area is also able to ensure market access for the increased crop production.

The Bank's sectoral organization handicaps cross-sector coordination.

Similar disconnects are seen in the financial sector. Respondents to the IEG survey noted that financial sector staff had been of little assistance in coming up with a realistic strategy for increasing access to financial services to support agricultural sector growth.

Similarly, agricultural education in universities is under the education sector; the agriculture sector does not have the main responsibility for it. Hence, there is little evidence of attempts to link support for technical education in agriculture with the needs of the agriculture sector. More than 80 percent of the survey respondents agreed that coordination between Bank staff in agriculture and in other sectors in the Africa Region is not good.

Eighty percent of staff survey respondents said that intersectoral coordination was not good.

Even within a country sector program, there can be little coordination between projects. IEG's project assessment report on the Seed System Development and the National Fertilizer Projects in Ethiopia (both approved in June 1995) found that a feature of the two projects was a lack of interlinkages and coordination in conceptualization, design, and implementation and among all parties involved (IEG 2007a, 2007b).

Despite going to the Bank's Board of Directors on the same date, each of the appraisal reports makes only minimal reference to the other project. Neither report considered how the two projects would harmonize their activities and there was little discussion of how they would engage with other activities—such as agricultural extension, research, and credit—that would be needed to ensure that the project activities increased agricultural productivity. The country review for Cameroon, drawing on earlier IEG

Even within a country's sector program, coordination may be poor.

work in the forest sector of that country (Essama-Nssah and Gockowski 2000), also found that an understanding of the multifaceted and interconnected nature of agriculture and the major role that low productivity in the sector has played in deforestation was missing from the country program.

How can knowledge about the interconnected nature of agricultural interventions best inform the design of future agriculture projects? A sectorwide approach, such as that used in the Zambia Agriculture Sector Investment Program, may not necessarily be the best answer if it leads to the design of complex projects. This may challenge limited country capacity. Past donor procedures have not been compatible with pooling resources, as was attempted initially in the Zambia intervention.

Once the overall menu of activities has been identified, separate interventions can also be undertaken, although given the sectoral organization of the Bank, they present a coordination challenge within the institution. The realization that agricultural development requires a multifaceted and coordinated approach has to flow through the different Bank teams working on different projects in one country.

The broadening of the rural agenda has been accompanied by reduced technical capacity in agriculture.

Beyond the Bank, as seen in chapter 2, when other donors are involved in the overall task of supporting agricultural development, donor coordination presents challenges in terms of agreeing on strategy and priorities. Programmatic and budgetary support lending, now on the increase in Africa, seem likely to make coordination more difficult. This is because the allocation of those funds rests with sector ministries, which are in most cases far less cooperative than the Bank's sector units.

Technical capacity within the Bank

Major reorganizations in the Bank have significantly reduced its technical capacity in agriculture.

Two major reorganizations in the Bank in the past 20 years have significantly reduced the Bank's technical capacity to support agriculture. In the 1960s and 1970s, considerable attention was given to technical aspects in components of agriculture projects. To do this, the Bank maintained a strong cadre of technical staff who came into the Bank in mid- to late career, and whose quality of support was acknowledged by the client countries and the world at large (World Bank 1991a). The major reorganization of the Bank in 1987 significantly reduced the number of agriculture technical staff in the Bank.[1] This was recognized as an issue by a study that examined aid to African agriculture in the late 1980s (World Bank 1991b).

A decade after the first reorganization, after the Bank was reorganized again, along matrix lines, the availability of technical staff eroded further. Analysis of data from the Bank's Human Resources Department found a considerable decline since 1997 in the number of technical staff (irrigation engineers and specialists in soils, extension, livestock, and other areas) mapped to ARD in the Africa Region. In 1997 there were 40 technical experts mapped to ARD in Africa, but in 2006 there were only 17 (appendix G).[2] More than 67 percent of the respondents to the IEG staff survey agreed that the Africa Region does not have an adequate level of technical staff skills to support implementation of agriculture projects.

The decline in the Bank's technical capacity happened partly because of the broadening of the rural agenda discussed in chapter 2. While social development, broad-based rural development, and other such concerns are important issues in rural space, the emphasis on those new areas in rural strategies has resulted in a staff of generalists rather than one of agricultural specialists. Human Resources data show that staff related to the newer agenda have increased from about 51 percent of staff working in ARD in 1997 to 71 percent in 2006. With such limited technical capacity in ARD, it has become difficult for the Bank to provide substantive direction and advice to countries on technical agricultural issues, especially since government sector staff with which the Bank interacts are still largely technical specialists. About 65 percent of the respondents to the IEG staff survey agreed or strongly agreed that the strategic approach by the Bank of focusing on rural development more broadly has diluted attention to technical issues in agriculture lending in Africa.

Decentralization also appears to have affected the Bank's capacity to support agricultural development.[3] The decentralization of Bank staff in the mid-1990s led to the increased hiring of local staff in the Bank's country offices. The decentralization improved understanding of country issues and reduced staff costs. While both are desirable goals, the tradeoff for this has been the reduced influence of internationally recruited staff with broad experience and knowledge of international good practices.

The decline in technical capacity appears to have affected the quality of the policy dialogue on agriculture with government ministries. More than 66 percent of the respondents to the IEG staff survey agreed that the policy dialogue bearing on rural development in the Africa Region does not adequately address technical issues in agricultural productivity (such as soil fertility, land management, land tenure, irrigation, and improved seeds).

The decline in capacity also has affected the quality of the agriculture lending program. Many country directors interviewed by ARD cited input and output marketing as areas where there were major problems in their countries, but they found that Bank staff were unable to help resolve these problems (World Bank 2005g). Other country directors noted that they could not get the skill mix they needed from rural staff for products such as Poverty Reduction Strategy Credits (PRSCs) and cited lack of attention by rural staff at the time of CAS formulation as a reason for smaller rural programs (World Bank 2005g).

Monitoring and evaluation

The Bank requires that each project approved have a monitoring and evaluation (M&E) system. Since January 1996, when Operations Policy and Country Services (OPCS) provided guidance to staff on preparing indicators, most projects have given increased attention to M&E. A review of 54 investment projects in the sample of 71 (see appendix A for how the 54 projects were selected) found that 73 percent of projects since 1996 have had agriculture-related indicators, compared with 27 percent during the period before 1996. Of all the projects that had indicators, most included output indicators, though the number of outcome and impact indicators has increased since 1996.

The loss of technical capacity has affected the quality of policy dialogue.

The types of output and outcome indicators in African agriculture projects vary widely, presumably reflecting the wide variation in project objectives and components. Project documents usually do not say how the indicators were selected, and the indicators listed are often not thoroughly defined. Though ARD is currently preparing guidance for designing indicators, no such guidance has existed up to now.

Even where there are indicators, the information reported in the completion reports is often of limited value for answering fundamental outcome and impact questions, such as who benefited, the development of which crops received support and how, and what gains can be attributed to the Bank project, among others. An internal ARD review in 2004 confirmed this finding.

The percentage of agriculture projects with agriculture-related indicators has grown considerably since the late 1990s.

A recent (February 2006) supervision mission for the Kenya Arid Lands Resource Management Project Phase II noted weaknesses in reporting, commenting that the reporting is overwhelmingly on activities undertaken, and not on their impact. IEG's assessment of the first Kenya Arid Lands Resource Management Project (IEG 2005b) had also found that only 5 of 19 indicators were impact indicators. In that report, IEG concluded that there could have been better assessment of qualitative aspects related to the responsiveness of district institutions and of poverty-focused activities, and whether benefits were being captured by the elite.

Where there are indicators, however, the information they provide is often of limited value.

Further, where there were indicators, and they were relevant, reporting was limited, often due to weak capacity in the country or because of weak or inadequate Bank supervision. Seventy-two percent of the completion reports reviewed mentioned problems with M&E that limited the ICR (Implementation Completion Report) team's ability to fully assess the project's outcome or impact. As a result, learning and scope for designing realistic follow-on interventions is limited.

The projects reviewed also tended to treat beneficiaries as undifferentiated groups.

Weakness in the M&E of agriculture projects in Africa has also been identified by an internal ARD study and by QAG. While weak M&E is not unique to the agriculture sector, since the outcomes of agriculture interventions are influenced by interventions in other sectors and by natural and other factors, it is critical that information on the Bank's activities be accurate. The review of project completion reports found several cases where weak M&E kept the ICR team from separating the project's contribution to the final outcome from external factors (such as weather events) or other projects that were implemented in the same period with similar objectives.

Though nearly half of food production in the Region is by women and issues affecting them are different, when a farmer is mentioned in project documents it is difficult to tell whether a male or female farmer is being discussed.

Another measurement issue common to the projects reviewed is the tendency to treat beneficiaries as undifferentiated groups. Few project documents provide a profile of the farmers that are expected to benefit. Instead, the typical document refers to beneficiaries in general terms such as *farmers, stakeholders,* or *smallholders.* It is important to differentiate among various farmer categories, as noted in recent sector work in Zambia (World Bank 2007e). On the potential to commercialize smallholders, the report says (pp. 7–8):

> *When considering the potential for smallholder commercialization, it is important to recognize that Zambian smallholders are not a homogeneous group of farmers. Understanding the heterogeneity of Zambia's rural households and their different potential as agricultural producers is critical to designing strategies for commercially viable smallholders. There are distinct differences in smallholder households' assets, human capital, income generating potential, and livelihood strategies.*

With the greater interest in promoting nontraditional export crops and with the increasing standards demanded by importers, understanding smallholder capacity will become even more important.

There is also a gender dimension to the issue of farmer profiles. Nearly 50 percent of food production in Africa is undertaken by women farmers, and the challenges they face in access to land, credit, and extension are different from those of their male counterparts. However, the portfolio review found that, in most cases, when a farmer is mentioned in project documents, it is difficult to tell whether a male or female farmer is being discussed. Only 2 of the 71 documents reviewed clearly link gender to the project objectives, include gender-specific subcomponents, and have indicators to measure the project's impact on women.

The literature also shows that changes in the division of labor occur over time for several reasons (Doss 1999). Men may move into activities that are traditionally the province of women when new opportunities arise and activities previously done by women become more productive or profitable (Doss 1999). This suggests the need for a more complete profile of the intended beneficiary households to effectively design Bank interventions that target farmers' needs and to report on variation in the impact of interventions on different beneficiaries.

The recent emphasis on client-responsive approaches to agricultural development requires even greater attention to the details of farm households. Project teams might argue that these details may be included in beneficiary surveys and other documents prepared for such interventions. However, those documents are not readily available, and since the information is not reported at the completion phase, it is not clear how much the information they contain contributes to learning or is a factor in assessing the Bank's contribution.

The portfolio review also found that 40 percent of the closed projects reported information on yield change through indicators, but the yields for each crop were reported in aggregate. This makes it impossible to discern differences among specific types of farming, production systems, or agro-ecological zones. In addition, the project information did not explain the criteria used to select the crops that were reported. In many cases, specific crops are also noted in the economic analysis section of project documents, but as with yield indictors, it is not clear whether these are the only crops supported by the project or why they were selected for the calculation of economic rate of return. This review concurs with the conclusion of an internal ARD review carried out in 2004, which noted that until the Bank addresses the insufficient use of outcome-oriented indicators, inadequate M&E, and reporting tools that are not designed to facilitate the description of project results, it is unlikely that the Bank will be able to effectively track the results of its interventions in the Africa Region in a meaningful way.

Country Factors

Without political will and commitment and capacity in the countries it supports, the Bank's activities on behalf of agricultural development are unlikely to be effective. Since there are potentially 47 borrower countries in the Region, it would be difficult to address country-specific issues. Instead, this section focuses on two broad factors across countries. These are critical aspects of the wider issue of governance, which evaluation findings and the literature indicate have affected the development of agriculture in Africa.[4] While political will and commitment and stability are less amenable to outside influence, the Bank can help build government capacity to formulate and implement sound policies through training and technical assistance programs.

Political will and commitment

Among the most important lessons for Africa from Asia's agricultural development experience is the necessity for political will at the highest levels. It translates directly into favorable policy environments and budget allocations to agricultural institutions and related infrastructure (IFPRI 2004b). With NEPAD and CAADP there now appears to be political commitment among African governments to support the development of their agriculture sectors. IEG's Poverty Reduction Strategy Paper (PRSP) review (IEG 2003b) also found that a large majority of PRSPs reviewed (94 percent) identified agricultural issues as central to the fight against poverty. African governments, many allocating less than 1 percent of their budget to agriculture, agreed in July 2003 at the Africa Union Summit to allocate at least 10 percent of national budgetary resources for implementation of policies and programs to support agricultural growth within five years. It remains to be seen whether the governments will be able to meet this commitment.

Political will and commitment for agricultural development appear to be growing in Africa.

Political commitment to develop agriculture was weak early in the post-independence era, as reflected in the budget support and the policy environment for the sector. Though there were variations in policies across countries, agriculture generally faced heavy taxation, and monopolistic parastatal marketing boards often fixed producer prices below market levels.

However, governments also transferred resources through input and credit subsidies. As a result, some analysts have argued that the governments in Africa followed a contradictory strategy, extracting surplus and transferring resources at the same time. Such an approach allowed the government to meet the needs of the smallholders and supply cheap food for the urban population. But the strategy was fiscally unsustainable and did not contribute to development of the agriculture sector.

Then, in the mid-1980s, African countries were faced with severe crises—alarming impoverishment, food shortages, low levels of literacy and health, a fall in commodity prices, and a stifling rise in the debt burden (IEG 1998a). Agricultural performance declined as the area under cultivation expanded and the best lands were exhausted. What followed was an era of structural adjustment reforms when, under pressure from the Bank and the IMF, several governments undertook major reforms.

Political commitment for both the sector reforms and agricultural development was often weak during the structural adjustment period. Many government decision makers did not accept the premise of the reforms and did not trust the working of the markets (Jayne and others 1999). Ministries of agriculture did not support a reduction in their functions, preferring to retain budgets and authority even where they or the central government made verbal commitments to liberalization (Foster and others 2001). Civil society organizations often opposed the reforms, arguing that they adversely affected the poor, and some expressed concern over losing sources of public revenue because of the reforms (Kherallah and others 2002).

That political commitment was weak is obvious in many ways.

In Senegal, for example, the government struggled to maintain control over the processing and marketing of groundnuts, its primary generator of export earnings (IFPRI 2000). In Mozambique, an IEG project assessment (IEG 2002a) found that it was widely believed in the country that a Bank-supported reform program to liberalize the cashew sector "killed" the economically viable cashew processing industry. While the findings of the Bank and the literature differ on this issue, and while there is recent evidence of increasing production using intermediate technologies, the reform process itself clearly contributed to poor relations between the Bank and the government. Overall, this appears to be a case of Bank conditions being pushed too far when a government was not convinced or committed. A National Bureau of Economic Research paper (NBER 2002, p. 28) argues:

> *The reforms took little note of important market imperfections. . . .There was virtually no attention paid to the credibility of policy changes and how to enhance it. The government made little effort to manage the political fallout that should have been quite predictable ex ante. And the World Bank did not sufficiently appreciate the ineffectiveness of buying reform through aid-cum-conditionality. In all these respects, Mozambican cashews provide an illuminating case study of the misfortunes that have befallen the reforms that African countries undertook in the last couple of decades.*

The lack of initial enthusiasm for policy reform by African leaders probably reflected doubts about how responsive the economy would be to these reforms (Jayne and others 1997). Hence, reforms were often undertaken because they were a condition of a Bank loan. The agriculture sector review for Kenya found that an important lesson from the experience of the Agricultural Sector Management, Parastatal Reform, and Economic and Public Sector Reform Projects was that relying largely on conditionality did not work, and that mechanisms were needed for Bank-client communication and greater consultation with politicians and civil society.

A review of completion reports of structural adjustment and investment operations in several countries points to a number of manifestations of weak political commitment—inadequate provision of counterpart funds for projects, delay in passing important regulations and in dismantling parastatals, inconsistent policy directives, and delay in adoption of policies, among others. A review of QAG's supervision assessments of agriculture projects confirmed that weak government commitment was a significant problem during Bank supervision. In 56 percent of the closed projects where borrower performance was rated unsatisfactory at project completion, weak political commitment was a factor (see box 4.1).

Weak commitment has contributed to underfunding of critical research and extension systems in several African countries. The review of the agriculture sector in Nigeria found that while the country has the largest agricultural

Box 4.1: Weak Political Commitment Has Been a Factor in Performance

The completion report of the Togo National Agriculture Services Project (fiscal 1998) rated borrower performance unsatisfactory and noted, "The government failed to provide, in a timely and adequate manner, its counterpart funding to implementing agencies, and, for reasons still unclear, it withdrew the extension services for cotton production from ICAT and returned them to SOTOCO from whom they had originally been transferred" (World Bank 2003f, p. 12).

The 1998 IEG study of the agriculture sector in Kenya (IEG 1998c) also found that lack of sufficient ownership was a weakness that had severely compromised developmental effectiveness of Bank-supported operations in Kenya.

research system in the Region, funding to the system was severely curtailed in the 1980s following the decline of oil prices. In Tanzania, IEG's recent assessment of the Second Agriculture Research Project (fiscal 1998) and the Second National Agricultural Extension Project (fiscal 1997) found that sustainability remains the biggest concern because there was insufficient attention to matching the scale of public sector activity to realistically projected resources.

In some cases the lack of political commitment reflected a deeper governance problem, and Bank project implementation did not always recognize this. The Cameroon, Kenya, and Tanzania agriculture sector reviews found that Bank interventions show little appreciation for the time required to carry out major reforms. In many cases, project completion reports and IEG project assessments also found that the pace and scope of the reform advocated in countries in the Region has been beyond the capabilities of the governments.

In Kenya, for example, reforms ran into political and implementation delays and reversals because of unrealistic expectations regarding the steps required for the reforms to pass through into legislation and implementation. In Zambia, the project assessment for the First and Second Privatization and Industrial Reform Credits found that the reform programs under the two projects were beyond the intent and capacity of the government to implement fully. In Cameroon and Nigeria, other factors played a part. The country sector reviews found that commitment to the agriculture sector rose and fell in tandem with oil revenues.

Many countries reversed reforms as a result of external shocks or changing economic conditions (IFPRI 2000). The civil war that erupted in northern and eastern Uganda, for example, forced the government to divert resources, or even ignore some of the components of liberalization such as a prohibition on printing currency to cover budget deficits (Bazaara 2001). Malawi reinstated fertilizer subsidies that were to be phased out in the mid-1980s because currency devaluation and the severance of transport routes through Mozambique significantly raised fertilizer prices (IFPRI 2000). In Ghana, the IEG project assessment report (IEG 2001) found that while structural adjustment was a major part of support to the country after 1990, in 1992, coinciding with elections, public expenditure financed by borrowing from the banking sector increased substantially when civil service salaries rose. This resulted in a large increase in the money supply and high inflation, and negated the reform principles.

Weak commitment also contributed to underfunding of critical research and extension systems.

External shocks or internal conditions caused many countries to reverse reforms.

Country capacity to support development of agriculture

Willingness and commitment are not by themselves enough to drive the development of agriculture. Capacity is also needed. In many African countries, weak capacity has prevented the state from effectively planning and budgeting, managing development assistance, and providing services (Commission for Africa 2005). In some countries, scientific and technically proficient

In many African countries, weak capacity has prevented effective planning and budgeting, management of development assistance, and provision of services.

staff are in short supply (Commission for Africa 2005). This problem is partly related to the quality of education in universities, which is not a subject of this review, but again reinforces the dependence of agricultural development on other sectors.

Enhancing the countries' institutional capacity has been high on the donor agenda for the past two decades and has also been an important aspect of many Bank agriculture projects. This study found Bank activities that provided training to support the establishment of early warning systems for droughts and other natural disasters, improve M&E capacity, develop information systems, and strengthen human resource capacity through higher education, among other pursuits. Bank projects have also provided support for the revitalization and restructuring of agricultural research capacity to improve its coherence and quality and for training of research staff.[5] The Bank has also provided support for CGIAR, which has invested more than $3.2 billion in nominal dollars in research and capacity strengthening in Africa since 1971 (see box 4.2). Further, since the Bank began to champion the cause of the developing countries in international trade agreements, strengthening their capacity to negotiate trade issues has also become part of the capacity-building agenda.

The Bank's contribution to policy formulation has picked up since 2002.

The Bank has also provided support for restructuring of line ministries and privatization of grain and agricultural marketing boards (an area critical to governance), developing management systems and capacity to improve the allocation and utilization of budgetary and manpower resources, and enhancing capacity to formulate rural and agricultural policies and strategies. Contribution to strategy formulation, in particular, has picked up since fiscal 2002.

Sometimes, privatization and restructuring of line ministries or parastatals was part of a larger economic reform program in the country, as in the case of the Zambia First and Second Privatization and Industrial Reforms Credits (fiscal 1992 and 1993). At other times, agriculture projects focused primarily on sector institutions, as in the Tanzania Agriculture Sector Management Project (fiscal 2004). The predominant emphasis of the Ethiopia Seed Sector Development Project (fiscal 1995) was also for institutional and human capacity building. The main activity was to restructure and decentralize the Ethiopian Seed Enterprise to create a commercially oriented agency.

Support for building the capacity of governments at the regional and subregional levels became important with the emphasis on decentralization in the Bank's client countries. In addition, several projects, particularly from the late 1990s, have provided support for strengthening producer organizations and farmer or user groups, either to take on more responsibility for operation and maintenance, as with water user groups, or to improve the negotiating capacity of producer organizations, as in the case of attempts to strengthen cotton or coffee producers. Since these projects have become popular, Bank projects have also attempted to provide training support to government officials to build their capacity to deliver cost-effective services to rural communities and producer organizations.

The Africa Region's self-evaluations and IEG project assessments show that the capacity-building aspect of the Bank's support has had much less success than anticipated. An IEG *Précis* reporting on capacity building in the agriculture sector in Africa found that "although some success has been achieved in implementing structural adjustment programs with a consequent reduction in government activities to a more manageable size and liberalization of economic policies that improved resource allocation and producer incentives, there has been less success in reviving the capacity of public institutions" (IEG 1999c, p. 2).

Even today, local agriculture ministries continue to be weak and relatively ineffective partners in promoting development of the agriculture sector. Weak borrower capacity was an important shortcoming in 77 percent of the cases where a Bank-supported intervention was rated unsatisfactory on outcome. This finding is of particular

Box 4.2: The History and Challenges of National Research Capacity in Africa

The CGIAR is a network of 15 international agricultural research centers, all of which currently have programs in Africa, though only 4 are located in the Region. In addition, African countries also have national agriculture research systems.

In the early 1960s, 10 percent of the agricultural researchers in Africa were African and 90 percent were expatriate staff. By the early 1990s, however, overseas training programs had reversed the ratio.

This massive capacity-building effort in Africa is an important success story, but knowledgeable observers generally agree that agricultural research in Africa today is weaker and at an earlier stage of institutional maturity than in Asia and Latin America. Currently, 80 percent (4,800) of Africa's agricultural researchers are concentrated in 13 countries, while the remaining 20 percent (1,200) are dispersed in 35 countries across the continent.

Soon after independence, many new governments nationalized or abolished the regional research institutes of their former colonial governments. The first 20 years after independence saw substantial growth in national agricultural research systems, but it was not systematically planned and came at some cost to research effectiveness. The process was fragmented as departments within different ministries built their own research capacity.

During the adjustment era in Africa, civil service reform provided the political space for consolidating and restructuring agricultural research across much of the continent. This period also saw renewed donor interest in funding national agricultural research in Africa. Consequently, research units, staff, and infrastructure were pulled out of the different ministries, especially ministries of agriculture and livestock, and consolidated under a single semi-autonomous structure.

By 1991, 28 of the 47 countries in Africa had adopted this structure for their national research programs. Donor funding facilitated this process but it also resulted in a shift to reliance on donor financing, as agricultural research lost its traditional budget within the line ministries at a time of budget stringency and reordering of government budgets. The effectiveness of research in Africa depends not only on the link between CGIAR institutions but also on the strength of the national research systems. The World Bank has been the largest supporter of national research systems. After more than 40 years of independence, however, many of those systems are weak and financially unstable. Promising reforms are under way in the systems of several countries and increased attention is being given to developing alliances and partnerships with universities and the private sector. The competitive grant programs have also grown rapidly.

In 1985 a group of donors at the Tokyo CGIAR meeting decided to create the Special Program for African Agricultural Research (SPAAR). The program was charged with improving the coordination of donor aid to agricultural research in Africa and helping strengthen the capacity of the national systems to use new technology from the CGIAR system. The decision to create the Forum for Agricultural Research in Africa (FARA) was made in 1997 during the Seventeenth Plenary Session of the SPAAR. FARA was envisioned as a facilitating and information exchange forum among the subregional organizations and as an apex body to represent Africa.

The activities of this partnership have not been independently assessed. However, the biggest challenge is that national systems today are dependent on donor support for a large amount of their resources, and it is not clear whether resources will be available in the future to sustain and strengthen their activities. This is also the challenge for local-level cooperative research ventures—for example, CGIAR institute collaborations such as that between CIAT and the Eastern and Central Africa Bean Research Network. The Bank has been supporting such networking but, given the persistent problem of financial sustainability in African agricultural research, there needs to be more aggressive targeting of, and support for, such regional cooperation options.

Sources: Eicher and Rukuni 2003; Ndiritu and others 2004; http://www.fara-africa.org/about-us/

concern, as Bank lending is shifting toward budgetary support, thrusting far more demand for management decision making in setting priorities on these weak ministries. In such projects, there is a need for realistic Bank analysis of current institutional capacity during project preparation and a clearly stated assessment of that capacity in the appraisal documents.

Unrealistic or overly ambitious project design has been a major factor and was a concern in almost half of the projects rated unsatisfactory on outcome. Several Bank projects have been unrealistic about the availability of resources to support activities after project completion. For example, the Tanzania Agriculture Sector Management Project (fiscal 1994) appraisal report had

The main factor undermining performance has been unrealistic or overly ambitious project design.

anticipated that the savings from the rationalization of the Ministry of Agriculture would finance incremental recurrent costs, but the recent project assessment found that the agriculture ministries appear just as short of operating funds now as they were before the project.

The Bank has a long record of such experience, particularly in Africa, as treasuries tend to take back savings from downsized ministries. Funding is generally closely related to staff numbers. A similar situation occurred in Kenya, where savings from Forest Department reforms were redirected elsewhere. It should be possible for the Bank, which does Public Expenditure Reviews, to subject sectoral projects to tougher appraisal standards with regard to their financial sustainability in light of known budgetary constraints (IEG 1999c).

Capacity enhancement has also been limited by weaknesses in training programs.

Success with capacity enhancement has also been limited by weaknesses in the training programs provided within Bank activities. In some cases relevant training was not organized; in other cases trained staff were not used effectively. A review by the Africa Region in 1997 found that project training is often the least-well-defined component of a project (World Bank 1997a). A review of completion reports found that in about 15 percent of the completed projects where outcome was rated unsatisfactory, weakness in training was identified as an issue. For example, the overall objective of the Mali Irrigation Promotion Project (fiscal 1997) was to improve and induce, through capacity-building activities, an expansion in small-scale irrigation, contributing to increased on-farm diversification of investments, productivity, and food security. The completion report for the project, while noting that the overall result of the capacity-building component was unsatisfactory, found that the training had been infrequent and insufficiently integrated into a plan suited to the training requirements of producers' organizations.

Too little attention has been given to incentives for retaining highly skilled technical staff.

The completion report of the Gambia Agricultural Services Project (fiscal 1993) found that staff trained abroad for the specific purpose of improving sectoral analytical capacity were reassigned to other departments. In Malawi, the impact of efforts to build capacity, especially in the Ministry of Agriculture, was limited because of the rapid turnover of counterpart staff, particularly in the economist streams (IEG 1998d). In other cases, such as the Mali Natural Resource Management Project (fiscal 1992), IEG found that trained government staff moved to the private sector, to NGOs, or even abroad once the project closed (IEG 2003e).

Inadequate attention has been given to incentives and other factors such as staff salaries and promotion incentives that are important for retaining highly skilled technical staff. It is now widely recognized that the underpayment of public servants is a source of capacity weakness throughout most of Africa and is a serious impediment to the effectiveness of capacity-building interventions (IEG 1999c). However, most of these factors cannot be adequately addressed in sector interventions and often need to be tackled through reform of government pay structures and performance assessment and reward systems.

Chapter 5

Evaluation Highlights

- The Bank has provided support in a large number of areas, some difficult to track, relevant to agricultural development.
- But those interventions have been scattered, and not linked together in a manner that recognizes the interconnected nature of agriculture activities.
- Weakness in the Bank data systems make it difficult to tell how much support has been provided in different areas.

A bowl of cassava. Photograph by R. Koenig, reproduced by permission of the photographer and Still Pictures.

The Bank's Contribution—A Thematic Assessment

Using the evidence presented in chapters 3 and 4, this chapter assesses the Bank's activities for their contribution to relieving the constraints on agricultural development as identified in chapter 2.

Agro-Ecological Diversity

The portfolio review found that the Bank has provided some support that has made research more responsive to the agro-ecological diversity of Africa—most national research systems now have zonal station responsibilities. However, there is little indication that Bank-supported projects beyond those involving research have adapted their activities to diverse agro-ecological conditions and production systems within countries.

Although the background discussion of project documents often refers to different agro-ecological zones, this is not followed through in the project description or linked to project activities.[1] The portfolio review found that documents for only 8 of the 71 sample projects incorporated specific activities related to the different agro-ecological conditions into the project design. Moreover, in most cases, there is little reporting on progress made in responding to the diverse agro-ecological conditions. For example, the Malawi Agricultural Services Project (fiscal 1993) had the development of technologies for different agro-ecological conditions as an objective, yet it is difficult to say whether the project succeeded in meeting that goal, because its design provided little information on how project gains would be assessed. The only reference to agro-ecological diversity in the project's completion report is that one activity carried out simple fertilizer trials and has resulted in the compilation of a database of region-specific fertilizer recommendations, but there is no discussion of how or if this information was used or transferred to farmers.

The Bank has helped make research more responsive to agro-ecological diversity.

Bank staff clearly recognize the importance of adapting to agro-ecological diversity but seem to have difficulty working it into project design. Project completion reports have identified the lack of attention to agro-ecological conditions as a factor in unsatisfactory performance. The completion report for the São Tomé and Principe Agricultural Privatization and Smallholder Development Project (fiscal 1992) attributed unsatisfactory Bank performance in part to the provision of seeds that were not adapted to the countries' agro-ecological diversity. Similarly, the completion report for the Sudan Emergency Drought Relief Project (fiscal 1992) noted that the project included forage varieties not suited to

But there is little indication that projects other than research have adapted their activities to diverse agro-ecological conditions within countries.

drought conditions, resulting in low germination rates.

The ability to respond to local conditions has been the primary appeal of projects that use decentralized or community-driven development approaches. Yet a review of the agriculture projects that are meant to be client-driven found little attempt even in these interventions to respond to agro-ecological diversity. For example, the appraisal document for the Ghana Community-Based Rural Development Project (fiscal 2005) does not respond to agro-ecological diversity in the country, nor does the Tanzania Agriculture Sector Development Project (fiscal 2006). More than 51 percent of the respondents to the IEG staff survey agreed that Bank agriculture projects in Africa are not able to respond adequately to the agro-ecological diversity and the needs of diverse production systems.

Fluctuating Rainfall and Droughts

Nineteen of the 262 component projects in Africa had drought-related activities.

Nineteen of the 262 agriculture-component projects in Africa supported activities related to droughts, according to IEG's recent Natural Disaster Study (IEG 2006c). Several other agriculture projects also supported activities that were expected to build country capacity to reduce the impact of emergencies (including those arising from pests and diseases). The activities included research and dissemination of drought-resistant seed varieties (Ghana Agriculture Research, fiscal 1991; Mali Agricultural Research, 1994; and Tanzania Agricultural Research, 1998). The CGIAR, with Bank support, has also made a major contribution in this area. Other activities include putting in place drought early warning systems, as in the Kenya Arid Lands Resource Management Project (fiscal 1996).

These activities may have reduced vulnerability, but their sustainability is questionable.

Though such activities may have helped reduce vulnerability, the poor sustainability record of Bank agriculture projects in Africa suggests that their long-run contribution to food security has been limited (see figure 3.2 in chapter 3). IEG reviews of completion reports have noted inadequate availability of resources to carry out activities beyond the Bank-supported projects or inadequate government commitment, among other things, as reasons. In the Zimbabwe Emergency Drought Relief Project (fiscal 1992), sustainability was a concern because of a lack of follow-through on a comprehensive policy and institutional framework for drought preparedness and drought mitigation. The Sudan Emergency Drought Recovery Project (fiscal 1992) was rated unlikely for sustainability based on its failure to generate the political support required for sustainable action on the formulation of food security policy.

IEG project assessments have also found sustainability to be a major concern. For example, the Kenya Arid Lands Resource Management Project was found to have established a drought early warning and response system, but of the 11 districts most adversely affected by drought, it said, "if there is no support from the broader government system in providing some resources, either through government's own resources or through mobilization of external assistance, the improvements introduced by the project will gradually atrophy" (IEG 2005b, p. 17).

The findings in the project assessments and the portfolio review show that Bank-supported agriculture activities in Africa have generally responded more to dealing with acute food insecurity when it occurs than to helping countries develop a long-term approach to address the factors that create food insecurity.[2] In the large areas of Africa where rainfall is highly variable, irrigation is extremely limited, and droughts are frequent, acute[3] and chronic food insecurity are inextricably linked. A drought worsens the situation of the millions who are chronically food insecure.

While Bank-supported activities have had some success with helping governments set up warning and drought management systems, sustainability is an issue even here. Achievements in dealing with chronic food insecurity have been poor. Despite its presence for more than two decades in several countries, Bank support has so far not been able to help countries increase agricultural productivity sufficiently to arrest declining per capita food availability. In most African countries, food insecurity is directly related to insufficient total food production, in contrast to South Asia and other Regions where food insecurity is primarily caused by poor distribution and lack of purchasing power (Sanchez 2002).

A very large percentage of drought-related project investment has been undertaken in response to emergencies. Of the 19 projects with activities that responded to droughts, the activities in 9 were solely for emergency mitigation. Of the other 10, several attempted to put in place long-term drought management. But this work was generally not aimed at improving agricultural productivity.

Where specific activities could lead to improved long-run productivity—research and dissemination of drought- and disease-resistant varieties, for example—results have been poor. Among the reasons for this poor performance are weak coordination between CGIAR research and Bank interventions,[4] inadequate extension, and farmer reluctance to adopt improved technologies because of a shortage of complementary inputs and credit. The last could have been addressed more aggressively had the Bank-supported activities shown an appreciation of the multifaceted nature of agricultural development.

In countries where droughts are very frequent, such as Malawi, the Bank's major response has been to provide emergency loans. About 80 percent of the credit for the Malawi Emergency Drought Recovery Project (fiscal 2003) was quick-disbursing assistance for agriculture inputs to provide immediate relief. The Bank has also provided technical support for mitigation and prevention of weather risk. However, it has provided only limited support through other agriculture projects in Malawi and has not been able to contribute much to increasing agricultural productivity.[5] Meanwhile, food security in the country has become more precarious.[6] The IEG assessment of the Malawi Emergency Drought Recovery Project acknowledges that "Over the past 10–15 years, Malawi has shifted from being a self-sufficient producer of maize in non-disaster years to being a regular net importer dependent on foreign assistance to achieve a national food balance" (IEG 2007f, p. 3). Though several factors have contributed to this change, inadequate attention to issues related to agricultural development has been a major factor. Agricultural growth in Malawi declined from 8.1 percent a year during 1990–2000 to 1.8 percent during 2000–04 (IEG 2007f).[7]

Projects have generally responded more to acute food insecurity than to dealing with its long-term causes.

In Ethiopia, where droughts and rainfall variability also contribute to food insecurity, the Bank has again not taken a strategic approach to reducing vulnerability. The 1995 assistance strategy for the country noted that the central objective was to reduce poverty. Improving the ability of the rural population to cope with periodic droughts and improve food security on a sustainable basis was meant to be a key element of this strategy.

Bank support has had some success in helping governments set up warning and management systems.

The assistance strategy also recognized that most Ethiopian agriculture is rain-fed, and highly variable rainfall and periodic drought create a high level of risk for farmers and uncertainty about the expected volume of domestic food production. These forces have produced a history of widespread famine that has exacted a devastating human toll. Steps to improve food security, including greater use of water resources in times of drought, are thus central to Ethiopia's development strategy.

However, during the period of review, there has been only very limited Bank lending to support development of irrigation (through a social fund) in a country that has so far developed only 170,000 hectares of its estimated irrigation potential of 2–3 million hectares (World Bank 2006a). A Food Security Project was approved in fiscal 2002; among other things, it was meant to focus on soil conservation and water harvesting. The project was expected to do this using a community-

Where droughts have been very frequent, the Bank's response has generally been to provide emergency loans.

driven approach. But for farmers to "demand" microprojects related to soil management, they must have appropriate knowledge about viable options with quick returns. This is rarely the case. The Ministry of Agriculture's midterm review report for the project (September 2006) shows that very few of the chosen subprojects were actually designed to improve land productivity.

The Bank has provided very limited lending for irrigation development.

Early findings from IEG's ongoing Ethiopia Country Assistance Evaluation suggest that the Bank's overall efforts in the agriculture sector have been disappointing. The 2003 CAS acknowledges, "Not only are poverty levels amongst the highest in the world, but the Ethiopian population is extremely vulnerable, especially because of its reliance on a rainfall-based economy. . . . While it is the drought that has sharply increased the numbers of affected people, underlying causes of vulnerability and related economic, social, and developmental deficiencies have to be addressed" (World Bank 2003e, p. 3).

The Bank has failed to take a long-term, strategic approach to drought and food security, in part because it has not taken a multifaceted approach to agricultural development. As a result, Bank support that could have led to major successes achieved much less than anticipated (see box 5.1 for an example).

A recent IEG review of CASs in 12 African countries where there were agriculture projects with drought components found that most of the discussion around food security involved the distribution of food aid (IEG 2006c). None of the CASs mentioned the role that sectors such as transport can play in increasing accessibility to drought-prone areas or decreasing their vulnerability. As a result, rural road development, which could make a major contribution to country capacity for drought management, is generally not part of a strategic drought management approach. Further, the portfolio review for the recent transport study (IEG 2007o) found that exposure to droughts was not a major factor in identifying the location of

Box 5.1: Bank Support for Fadama Project I in Nigeria: Achievements Constrained by Lack of a Multifaceted Approach

The Bank's support of fadama irrigation in Nigeria attempts to make agricultural production less dependent on erratic rainfall. Fadama lands are flood plains and low-lying areas over shallow aquifers along Nigeria's river system. The farmers of northern Nigeria have long used water drawn from shallow wells or streams to irrigate fadama lands, where they cultivate small areas during the dry season.

A pilot initiative financed by the Bank under the National Agricultural Development Projects (ADPs), undertaken prior to the study period for this review, helped introduce low-cost tubewell drilling and irrigation by pump in the traditional fadama farming areas. The fadama components were the most successful elements of the ADPs and were scaled up into a free-standing project, the first National Fadama Development Project (Fadama I, fiscal 1993).

Fadama I raised crop yields, but profits were low because farmers lacked access to markets and insufficient attention had been given to downstream processing and marketing. Achievements were also constrained by land tenure uncertainties, which exacerbated traditional tensions between farmers and pastoralists.

Fadama II (fiscal 2004) attempts to address some of the shortcomings using a community-driven development (CDD) approach. It also proposes to support demand-driven research and extension, and better access to inputs and markets. While it is expected to handle some of the challenges that constrained agriculture production in the first project, it is not clear that it will do so. Previous IEG assessments of CDD projects have often found that such projects are unable to give adequate attention to sector-specific technical issues.

It is too soon to tell whether Fadama II will succeed. The latest supervision report notes that few of the subprojects have started yielding benefits to communities. There are also concerns about inadequate maintenance plans and insufficient capacity of facilitators and private service providers to provide adequate technical support to farmers. While the Bank can be credited for having stayed for the long haul in fadama areas, inadequate recognition of the multifaceted nature of agriculture has restrained its achievements.

rural roads in Africa. In fact, there is very little information in project completion reports on how the locations for particular rural roads were selected.

The portfolio review for this study found that although several projects have dealt with both food security and drought, few adequately addressed the causative links between the two. While improving food security was a stated objective of 8 of the 71 projects in the sample, only 1 of the 8 specifically links the issue of food security to drought, despite the frequent recurrence of droughts in the Africa Region.[8]

Partly because of the weaknesses in analytic work already noted, the Bank's project appraisal documents do not show an appreciation of the diversified coping strategies that traditionally have been followed to minimize risks of food insecurity. In diversified cropping systems, some crops, such as cassava and millet, have been particularly important because of their drought resilience. Figure 5.1 shows cassava yields in comparison with maize yields in drought years. While the Bank has contributed to development of improved millet and cassava varieties, the assessment did not find much evidence that Bank projects had a long-term strategic approach to linking the development of cassava or millet to building food security in individual countries or building on the resilience of a traditional system with built-in security measures.

Although several projects have dealt with food security and drought, few adequately addressed the causative links between them.

Contributing effectively to cassava development seems to have been a missed opportunity for the Bank (box 5.2).[9] A sound poverty focus in circumstances of declining soil fertility and high input costs in many African countries would likely point toward directing more effort to the development of crops of particular importance for the poor.

Similarly, although there are projects that support livestock development activities, there are few (the Kenya Arid Land Project may be the only exception) that recognize the value of livestock in the diversified production systems farmers use to cope with drought and that attempt to increase the efficiency of livestock production.

Soil Fertility

A review of CASs and project documents shows that the Bank does not appear to have engaged its African clients in serious policy dialogue about

Figure 5.1: Production of Maize and Cassava in Six Drought-Affected Countries of Southern Africa

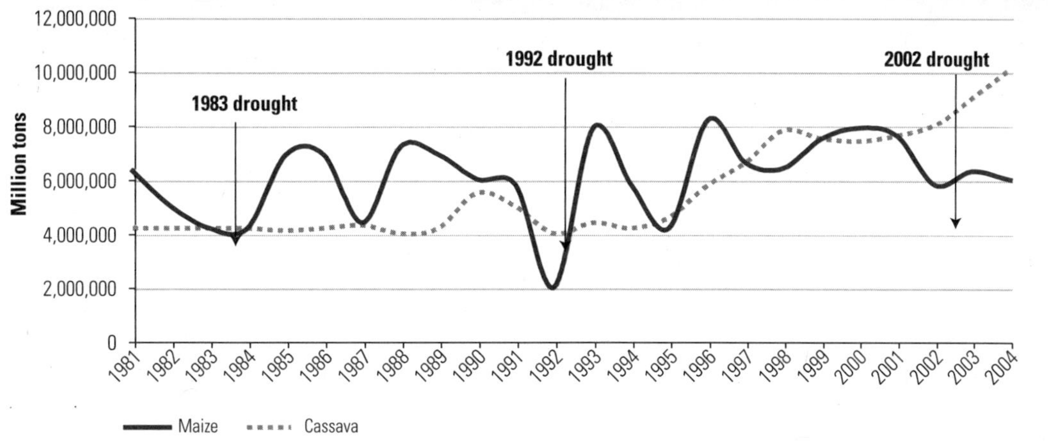

Source: FAO Web site 2005.

Note: Countries include Lesotho, Malawi, Mozambique, Swaziland, Zambia, and Zimbabwe. The sensitivity of maize to moisture variations is compared with "cassava," representing roots and tubers.

Box 5.2: Cassava: A Missed Opportunity for the Bank to Contribute to Food Security

Cassava is Africa's second-most important food staple based on per capita calories consumed, and the Region produces half of the world's supply of the staple. Cassava provides a reliable source of food during drought (due to flexibility of harvesting), locust attacks, and the *hungry season*—the period before seasonal food crops are ready for harvest.

Cassava is grown in about 40 African countries by millions of poor farmers, many of them women, often on marginal land. Though estimates differ, about 70 percent of Africa's cassava output is harvested in Nigeria, where a number of factors have come together to allow its successful transformation from a low-yielding subsistence crop to a high-yielding crop produced primarily for urban markets. Availability of improved and disease-resistant varieties was only one of those factors (see appendix L).

African policy makers and most donor agencies neglected cassava for numerous reasons (FAO and IFAD 2005) until the late 1980s, when the Rockefeller Foundation initiated a Collaborative Study of Cassava in Africa. Then, in the mid-1990s, the FAO formally recognized the importance of cassava as a food security crop. This was followed by the Global Cassava Development Strategy (GCDS), an initiative spearheaded by the FAO and IFAD and formalized in 2002 for identifying opportunities and constraints to cassava production and processing. The strategy provides a framework for technical cooperation in research and technology transfer and for future debates on global issues affecting cassava. NEPAD has also identified cassava as a poverty fighter (NEPAD 2004; Whingwiri 2004) and has developed a market-orientated strategy to develop the commodity, which is based on the GCDS.

Where has the Bank been?

CGIAR institutions and Bank-supported research projects have contributed to the development of improved varieties and disease/pest control for cassava, but the linkages between CGIAR research and Bank projects have been weak. Country factors clearly played the key role in the cassava transformation in Nigeria, and the Bank appears to have had a minimal role.

Between 1993 and 1999 the Bank did not approve any new projects in Nigeria because of governance problems. Nor did it support analytical work that could help build the basis for future agriculture support in this area. Current Bank analytical work for Nigeria does not even show an adequate appreciation of the reasons for the increase in production of cassava (appendix L).

The Cameroon, Kenya, and Tanzania country reviews also found a lack of appreciation in the Bank's strategy statements and activities for the important role of cassava and other root crops in providing food security. The portfolio review shows that despite the recognition of the importance of the crop in the local farming system in Bank project documents, projects have not taken a strategic approach to building on its strength as a food security measure.

The Bank as an organization is not even a member of the FAO and IFAD initiative on GCDS. It is not clear why this is so. However, many Western food policy analysts still consider cassava an inferior food whose per capita consumption is expected to decline with increasing per capita incomes, and it is possible that the Bank approach has been influenced by this thinking. Given the dramatic increase in its production and use in Africa and its role in food security, it is clear that a decline in consumption of that crop is not likely in the near future.

Sources: http://www.fao.org/ag/AGP/AGPC/gcds/GCS.htm (March 21, 2007); FAO and IFAD 2005; study research.

the Region's declining soil fertility. About 68 percent of the CASs reviewed did not mention soil fertility. The portfolio review also found that less than 10 percent of the project documents discuss the inherent limitations of African soils. While the appraisal documents for 27 of the 71 projects in the portfolio review do refer to declining soil fertility, most (25) do not recognize the centrality of the problem to agricultural development. This review found that in most cases soil fertility was "tacked on" as an issue in the project. For example, the objective of the Tanzania National Agricultural Extension Project Phase II (fiscal 1997) was to "continue to improve the delivery of extension services to smallholder farmers for increasing their incomes and productivity, while improving its relevance, sustainability, and cost-effectiveness," but the appraisal states that "through the dissemination of messages related to improved fallow, afforestation, and anti-erosion techniques, the project would also have a positive impact on soil fertility, conservation, and water management" (World Bank 2006l, p. 14).[10]

There has been limited policy dialogue about the Region's declining soil fertility.

The Bank has often taken the lead in engaging its clients and the international community in discussion and debate on issues of such global and regional importance as soil fertility. So it is surprising that the issue has received so little attention. Bank staff are aware of the importance of the issue, however, and the CGIAR research institutions have been identifying and testing new soil management practices for some time. More recently, the Bank became party to the *Terr Africa Regional Initiative.* Launched in 2005, this is a multidimensional partnership that is expected to promote a collective approach to sustainable land management in Africa. The Bank has even supported the *Soil Fertility Initiative*[11] in Africa, but has not followed through on the initiative with either a serious policy dialogue with its clients or substantive funding support.

Interviews of Bank agriculture staff in the Africa Region and in ARD revealed several issues that may have contributed to the neglect of soil fertility. Among these are IDA funding constraints, shortage of technical staff, a sense among Bank management and staff that it would be another add-on among too many others, and an impression that this is mainly an FAO agenda.

The Bank appears to have seen soil fertility more as an environmental than an agricultural productivity issue. The portfolio review found that where project documents discuss soil fertility, the emphasis is more on halting land degradation and the consequent environmental damage than on directly addressing the link between declining soil fertility and agricultural development.[12, 13] This appears to have happened partly because environmental conservation became a priority within the Bank following the United Nations Conference on Environment and Development in Rio de Janeiro in 1992.

Concern about declining soil fertility has now led some countries, such as Malawi and Tanzania, to reinstate fertilizer subsidies, a common policy in earlier periods, as discussed in chapter 2. Many African governments and some donors believe that some food security and environmental issues could be addressed by input subsidies (Kelly, Adesina, and Gordon 2003).[14] A leading proponent of fertilizer subsidies has advocated large-scale distribution of low-cost or no-cost fertilizer as a way to help African smallholders escape the poverty trap (World Bank 2007a). Given Africa's current precarious position—rapidly declining soil fertility, very high fertilizer prices, and no easy short-term way of bringing them down to a reasonable level—it is creditable that the Bank has begun exploring ways of making fertilizers affordable for poor farmers. The above-mentioned Bank sector work (World Bank 2007a) summarizes some lessons learned and guidelines for increasing access to fertilizers by smallholders in Africa, but it is not clear how far the recommendations are being incorporated in Bank lending. In this connection, it will also be worth exploring how Kenya has succeeded in experiencing a tremendous growth in fertilizer use, as is evident in the literature (Ariga, Jayne, and Nyoro 2006), in the context of the forthcoming IEG agriculture study.

The Bank appears to have seen soil fertility more as an environmental than an agricultural productivity issue.

Water

The Bank supported 31 projects with irrigation components in the Region during fiscal 1991–2006. There have been few free-standing irrigation projects, and in only 8 of them was the irrigation component 45 percent or more. The largest share of the total lending of $343.5 for irrigation in Africa went to Mali (17 percent) in four projects, followed by Madagascar (11 percent).

There have been only a few free-standing irrigation projects.

The Pilot Private Irrigation Promotion Project (fiscal 1997) in Mali was expected to enhance the capacity of private institutions involved in providing equipment, services, and financing for small-scale irrigation investments. The findings of IEG's recent assessment indicate that the project failed to achieve its objective. It was expected that on-farm investments, induced by the project's technical assistance efforts, would lead to the rehabilitation of about 400 hectares and to the establishment of about 600 hectares of newly irrigated land. However, only 10 hectares were rehabilitated and no investments were made for new small-scale irrigation schemes, and the impact on private sector development was insignificant (IEG 2007i).

Concern about declining soil fertility has led some countries to reinstate fertilizer subsidies.

The Bank has also helped promote private sector development in irrigation.

The Bank has also helped promote private sector development in irrigation in countries such as Burkina Faso and Niger in West Africa. Among other things, the pilot projects supported services for on-farm demonstrations of small-scale irrigation equipment and techniques; promotion of markets for small-scale irrigation products, inputs, and services; and facilitation of access to financial services. It appears that these have had reasonable success, although neither project has been independently evaluated.

Current databases make it difficult to identify Bank support for water management in rain-fed areas.

In Madagascar, one of the two countries in Africa that have the maximum area under irrigation, the Bank provided support for irrigation development through three projects for a total of $37.95 million. A fourth project, on Irrigation and Watershed Management, was approved in November 2006. The Bank's support in Madagascar has largely been for institutional reform, specifically privatization of public and parastatal irrigation organizations in the early 1990s, and support for improved operation and maintenance (O&M), partly through transfer of the management of irrigation schemes to water-user associations.

The latest project aims to adopt a contractual approach that empowers stakeholders and clarifies their roles. Although it is too early to comment on project performance, a recent IEG mission to Madagascar found that transport and market access are major constraints to inputs and outputs, as is a dearth of agricultural credit on the appropriate scale. Experience from other countries in Africa shows that lack of attention to these factors has often constrained the achievements of Bank irrigation projects.

Forty-one percent of the portfolio had seed-related activities.

In some other countries, including Ghana and Nigeria, the Bank supported an irrigation component in a CDD operation or a social fund (the Community-Based Rural Development Project in Ghana, the Second National Fadama Project in Nigeria, and the Social Fund Project in Ethiopia). IEG's evaluation of community-based and community-driven approaches (IEG 2005a) noted the problem of sustainability of subprojects constructed under these interventions because of the lack of local community capacity and resources for O&M. The Ghana and the Nigeria projects are still being implemented, but the Ethiopia project has closed and the completion report itself rates sustainability unlikely.

Bank support for water management in rain-fed areas is difficult to identify because there is no system to track such projects. By looking at specific water resource management and environment "theme" codes (see appendix A), this review was able to find several interventions with small subcomponents for improved natural resource management.

The identified projects have attempted water harvesting and management (for example, Mauritania Rain-fed Natural Resource Management [fiscal 1997] and Madagascar Environment II [fiscal 1997]). In most of these interventions, physical targets are achieved or exceeded, but the projects themselves have not been sufficiently integrated with the countries' agricultural development strategy. Further, M&E has been very weak, so it is difficult to assess what has worked and what has not. The literature, however, suggests that such small-scale, technically simple water management systems can be effective in rain-fed areas (Sasakawa Africa Association 2004a; IFPRI 2005a).

Seeds

The Bank's database does not track projects designed to contribute to the production, distribution, and promotion of improved seeds. To identify such projects, IEG relied on information from the portfolio review. Forty-one percent of the portfolio was found to have seed-related activities. Most were investment projects, but there were also a handful of adjustment credits that sought to liberalize seed production and marketing or to develop a policy framework for market-based seed distribution. The latter also emphasized involvement of the private sector in input delivery systems.

The development of new seed varieties is mostly attributable to the work of the CGIAR, which the Bank supports.[15] However, Bank projects have provided opportunities for testing and scaling up of technologies developed elsewhere, particularly for crops such as maize. Among the activities supported by the projects identified by the portfolio review were research and dissemination of improved varieties, seed multiplication and production, provision of seeds in response to an emergency or as part of a safety net, and improving seed quality through construction of storage facilities or quality inspection services. The Togo National Agriculture Service Project (fiscal 1998), for example, was to support the production of seed for the major crops cultivated in the country. The Ethiopia National Fertilizer Sector Project (fiscal 1995) was to support the generation and dissemination of improved technology packages (including seeds).

Although the Bank, CGIAR, and other donors have worked on the development and distribution of improved seed varieties, the evidence in the literature suggests that the number of farmers regularly using that seed remains small (Kelly, Adesina, and Gordon 2003; Maredia and Raitzer 2006). The total area of Sub-Saharan Africa planted with improved varieties developed by CGIAR for 10 major food crops was about 11 percent of the total planted area in the late 1990s, compared with 55 percent in Asia, 30 percent in Latin America, and 48 percent in the Middle East and North Africa (Maredia and Raitzer 2006). Documented yield effects are variable across crops. Evenson (2003) estimated CGIAR contributions to yield growth based on genetic improvement in African crops to be in the range of 0.11 to 0.13 percent per year. This is significantly below the annual average yield growth of 0.30 to 0.33 percent across all developing regions.

A critical weakness in several countries has been the lack of seed multiplication capacity. Over the past two decades, most governments in the Africa Region closed their public seed companies in the belief that the private sector would take over. However, this has happened in only a few countries in areas with relatively good infrastructure and for only a few crops, such as hybrid maize, where profit margins are relatively large. Bank projects have not been very successful in promoting private sector participation in seed production for most other crops.

Constraints on seed production capacity have also been an issue in some countries, as IEG noted in its assessment of the Ghana Agricultural Extension Project (fiscal 1992; IEG 2001). In Ghana and elsewhere, the government's inability to establish transparent conditions for entry have made it difficult for the private sector to participate.

Development of new seed varieties is mostly attributable to the CGIAR, but the Bank has provided opportunities for testing and scaling up.

The only free-standing seed project in the portfolio, the Ethiopia Seed Development Project (fiscal 1995), made little progress toward the government's objective of privatizing the seed sector. Informal seed production by farmers did not develop, private wholesalers and retailers left the market, and no new private businesses entered the market. While the project attempted to lay the foundation for a competitive seed industry, the public and private sectors remained unequal competitors. In this approach, the government agricultural extension service was provided with seed and fertilizer. Farmers also had access to credit to buy seeds from the government. The same facility was not available for seeds bought from the private sector. This limited the demand for seeds from private entrepreneurs, who left the market (IEG 2007b).

But the number of farmers using those seeds remains small.

Seed multiplication and production capacity have been issues in some countries.

The literature also shows that most countries in Africa have a variety of registration and certification regulations to protect farmers against purchase of poor-quality seeds. However, the high cost in getting approvals, together with the small size of seed markets, has been a disincentive to the private sector (Poulton and others 2006). More recently, projects such as the Tanzania Participatory Agriculture Development and Empowerment Project (fiscal 2003) have begun involving farmer groups in the production and distribution of improved varieties (Anderson and others 2005). However, the project is still being implemented. Whether these interventions can help set in place sustainable multiplication and distribution systems remains to be seen.

One factor that contributes to farmer reluctance to use improved seeds is the affordability of fertilizers. Research in Malawi has shown that farmers have not adopted hybrid seeds even when they are available because of the high cost of fertilizers (Peters 2002).[16] Women farmers find it even more difficult to buy fertilizers because they do not usually have access to money from the sale of cash crops (Gladwin 2002).

Affordability of fertilizers and lack of cash or credit limit seed use by farmers.

Another factor affecting the use of improved seeds is the credit or cash needed to purchase them. Traditionally the seeds used by African farmers have been collected at the end of a cropping season and saved on farms. With hybrids, particularly for crops such as maize, farmers have to purchase new seeds each year, but few have the cash or access to credit for such purchases. While various attempts have been made to improve the affordability constraint in countries such as Zimbabwe by supplying seeds in small packets (Kelly, Adesina, and Gordon 2003), these have not been adequate.

Farmer willingness to buy is affected by perceptions of risk.

The vulnerability of hybrid varieties of several crops to diseases and pests has also been found to be a problem.[17, 18] Given the fragile environment and the risk aversion of the average African farmers, their willingness to buy inputs even if they are available in the market also depends on whether they expect to get a good price for what they sell.[19] The experience with maize in Africa shows that small farmers use improved seeds and complementary inputs if the technology, infrastructure, and overall macroeconomic environment are appropriate (IFPRI 2005b). Weakness in extension can also be a significant handicap.[20] Hence, the availability of improved seeds alone is not enough to increase yields.

The Bank's support to address the major financial constraint for farmers has been limited.

Credit and Rural Finance

With the Bank's existing coding system, it has been difficult to get a complete picture of the institution's support for activities in this area. It was possible for this review to cross-check the codes for "banking," "general finance," and "microfinance" in the Bank's database against the 262 projects with agriculture components. Through this analysis, IEG found that 38 of the 262 projects in Africa, 14 percent of the portfolio, had some aspect of agricultural credit and financial services, though there are no free-standing credit projects among the 262 projects.

There are very few investment operations among the 38 projects identified that have attempted to address the credit constraint of smallholders. Two examples include the Ethiopia Fertilizer Project (fiscal 1995) and the Rwanda Agricultural and Rural Market Development Project (fiscal 2000). The Rwanda project provides for farmer access to cooperative credit for input acquisition. More recently, the Mali Agricultural Competitiveness and Diversification Project (fiscal 2006) aims to facilitate access to capital and financial services for the private actors involved in the agricultural supply chains.

Other projects, such as the Guinea National Agricultural Export Promotion Project (fiscal 1993) and the Lesotho Industry and Agro-industry Project (fiscal 1991), have attempted to ease the financial constraint of farmers growing export crops. The Lesotho project, for example, was to encourage foreign and indigenous investment in the industrial and agro-industries sectors.

Where credit and financial services were part of a structural adjustment intervention, the focus was primarily on improving the overall enabling environment for development of a healthy financial sector.

Projects such as Ghana Rural Financial Services and Benin Rural Savings have provided support for rural credit as a part of the financial systems approach.[21] The Ghana project, which is still active, seeks to promote growth and reduce poverty in Ghana by broadening financial intermediation in rural areas. However, it has not provided support for agriculture development.[22]

In the Bank's data system, both the Ghana and the Benin projects lack agriculture codes. Though some may consider this mainly a coding issue, given the sectoral nature of the institution (as discussed in chapter 4), this can easily lead to lack of coordination of the activities of these interventions with other Bank-supported activities in the agriculture sector, a major concern of this review.

Past IEG studies have noted the low and declining level of support from the Bank, particularly for rural credit in Africa.[23] One reason for the low level of support is the weak performance of interventions in this area, as demonstrated by a review of completion reports and the findings of an IEG study of lines of credit (IEG 2006h). Weakness in performance of credit components can be attributed to weak implementation of Bank guidelines, particularly regarding eligibility and performance of financial intermediaries; lack of adequate Bank follow-through on reforms implemented; inadequate government ownership of the reform process; and the weak macro environment to support viable financial institutions, among other things.

An ARD review of rural finance activities noted a Bank-wide decline in credit lines and an increase in grant support. It may not be a bad thing that there are now more grants—it could be an appropriate response to the many obstacles involved to establishing a robust and sustainable rural credit system in many countries. There may be room for both grants and credit in the Bank's toolkit, and all options should continue to be explored for the most appropriate way to provide farmers with the necessary means of increasing productivity and incomes.

Weak past performance does not mean that the Bank cannot support activities well in this area. As noted in chapter 2, CGAP (to which the Bank contributes) research has made a contribution toward identifying viable and sustainable modalities for providing agricultural credit to farmers, which may help overcome the challenges identified in box 2.1. IEG's lines of credit study found that the demand from governments remains strong in this area. The study notes that "LOC [line of credit] can be a useful instrument when used well, and despite generally poor designs and outcomes, should not be entirely discarded from the Bank's lending toolkits" (IEG 2006h, pp. 32–33). However, the need is for the Bank to take greater care in designing and supervising these operations and to consistently follow Bank guidelines.

Credit components have performed below par.

There may be room for both grants and credit in the Bank's toolkit.

A 1996 IEG review of agriculture credit also suggested that subsidies could be appropriate under certain conditions, and the Bank committed itself at that time to calculate a subsidy dependence index for all rural lines of credit. In spite of that commitment, the IEG 2006 lines of credit study found that the Bank rarely undertook an analysis of the subsidy, indicating that there is little transparency with respect to subsidies in Bank operations.

Transport Infrastructure

Projects with agriculture components have made only a limited contribution to improving transport infrastructure for market access.[24] An examination of the investment in transport infrastructure in the 262 projects found only 54 with transport infrastructure components and a total of $634.1 million spent on those components over a period of 15 years. In the other 208 projects (nearly 80 percent of the projects with agriculture components), no investment was made in transport infrastructure. It could be argued that transport projects in the same area as the 208 agriculture projects might have helped improve farmers' access to markets. However, given the sectoral organization of the Bank, and the limited coordination among the Bank's various sectoral units and government ministries, there is no evidence that the process of selecting rural roads in transport projects is part of a deliberate, coordinated approach to developing agriculture.

Projects with agriculture components have made only a limited contribution to addressing transportation needs for agricultural development.

IEG's recently completed evaluation of the transport sector (IEG 2007o) found that no impact evaluations had been carried out in the Africa Region for transport interventions, which makes it very difficult to say anything about the contribution of these interventions to agricultural development. In addition, 80 percent of the respondents to the IEG staff survey agreed that coordination between Bank staff working on agriculture and those working in other sectors in the Africa Region is not good.

Even where there was investment in transport as a part of an agriculture project it was usually not done as a part of a multifaceted approach to agricultural development in the country. Of the 54 projects with transport infrastructure components, 18 are either structural or sector adjustments or economic recovery loans. The structural adjustments had some features associated with regulatory, institutional, and management reforms in the transport sector. When such reforms sought to reduce transport costs and improve services, as in Cameroon Structural Adjustment III, they likely provided an indirect stimulus to agricultural activities. However, this was not attempted as a part of a strategic approach for agricultural development.

Where there were investments in transport, they were not undertaken as part of a multifaceted approach to the development of agriculture.

In the emergency recovery loans the goal was to respond to the emergency rather than to address the longer-term development of agriculture. For example, in the Emergency Reconstruction Project (fiscal 2001) in Eritrea, even though 20 percent of the credit amount was for transport, it was primarily for the restoration or rehabilitation of key roads and bridges damaged by the war, restoration and provision of access to settlements and camps, and provision of improved access roads to areas of recurrent drought and famine.

A large percentage of the remaining investments in transport infrastructure are through interventions that finance demand-driven community-based infrastructure, such as community roads and bridges. Examples include the Malawi Social Action Fund Project (fiscal 2003), the Nigeria Local Empowerment and Environment Management Project (fiscal 2004), and the Mali Rural Infrastructure Project (fiscal 2000). Strategic development of the agriculture sector is not the objective of these interventions, which are primarily aimed at building local capacity and providing communities with access to social and economic infrastructure.

Some investments in the early part of the review period, such as the Agricultural Services Project (1992) in the Central African Republic, attempted to increase the road network as a strategy to improve productivity in rural areas. However, the achievements of the aforementioned project were limited, because its implementation was adversely affected by civil unrest.

Some other recent projects have been designed specifically to improve farmers' access to markets by road. The Mali Agriculture Competitiveness and Diversification Project (fiscal 2006) is an example of a project that is attempting to improve the performance of supply chains for a range of agricultural, livestock, fishery, and gathering products, for which Mali has a strong comparative advantage. The project proposes to improve rural roads for the collection of cotton and other agricultural produce. Another example is the Zambia Agriculture Development Sector Program (fiscal 2006), which aims to support increased commercialization of smallholder agriculture through improved productivity, quality, and efficiency of value chains where smallholders participate. The project will provide resources to rehabilitate and maintain feeder and district roads of economic importance in areas with high agricultural potential. It is too early to comment on the performance of these interventions.

Given the small size of several countries in Africa, regional programs can be very important to ensuring adequate transport coverage. The Bank has supported a regional program on Africa Transport Policy to improve transport sector performance by promoting policy reform and institutional changes in 32 countries in the Region. A recent IEG review of regional programs

(IEG 2007n) found that the program has made an important contribution to transport sector–level knowledge and expertise. The Bank is also supporting other infrastructure-related regional interventions, such as the Africa Trade and Transport Facility (fiscal 2006). It is too early to say how these projects will affect development of agriculture in the countries.

Extension

The Bank's approach to extension changed over the period 1991–2006.[25] Before the training and visit (T&V) approach fell out of favor in the late 1990s, the World Bank provided substantial financial support for this approach in several African countries.

The T&V approach provided extension services to farmers using trained public extension agents. One of the major concerns with the approach was the inability of government to meet the large recurrent cost on project completion. In the early years, T&V was also "top-down" and lacked systematic farmer participation, although this constraint was partly overcome in later years. Bank extension projects approved in the late 1990s increasingly provided for greater farmer participation, as in the Tanzania National Agricultural Extension II (fiscal 1997) and Burkina Faso National Agricultural Services Development II (fiscal 1998) Projects.

Beyond farmer participation, during the 1990s there was greater interest in promoting alternative extension concepts, with stronger participatory aspects, greater pluralism, and smaller public organizations (Anderson, Feder, and Ganguly 2006). Appendix H provides examples of alternative service providers that have become popular.

Based on country reviews and project assessments, IEG finds that the Bank's borrowers in the Africa Region appreciate the important role of technology dissemination in increasing productivity in agriculture. This is in agreement with the findings of earlier IEG work (IEG 1997b).

Extending knowledge will undoubtedly continue to play an extremely important role in the development of agriculture in Africa. For example, improved techniques can help address the large gap between potential and actual crop yields. The InterAcademy Council Report (2004)[26] found that gaps in yield within Africa are far greater than the gaps cited between Africa and the rest of the world. The report also found that "technology already 'on the shelf' has the potential to enhance land productivity in Africa once adapted and fine-tuned to location specific situations" (p. 75).

The Bank's borrowers in Africa appreciate the important role of extension in increasing productivity.

Extending knowledge can also improve management practices—with dramatic results. When the right varieties and good crop management techniques are used in combination, less fertilizer produces a higher yield. Heerink (2005) notes that only about 30 percent of the nitrogen from fertilizers is used by crops in West Africa. However, the benefits to households' food security from increased use of fertilizer and hybrid seed are unlikely to be fully realized without improvements in the efficiency of fertilizer use (Orr 2000). Timing and method of fertilizer application are significant problems that also can be addressed with good extension.

Extension can help improve crop yields and management practices.

Farmers can also improve water management if they have access to improved practices. In the Sahel, only 10–15 percent of rainwater is used for plant growth, and the remainder is lost through run-off, soil evaporation, or drainage (Heerink 2005). Farmers need information on existing low-cost, low-capital technologies for water harvesting. Knowledge transfer can also be important for the rehabilitation and maintenance of existing irrigation infrastructure.

Despite all the demand-driven and partnership approaches that the Bank has supported since it abandoned T&V, a viable and sustainable option to replace T&V has yet to be developed for Africa. Apart from a range of combinations of pluralistic approaches, some including Farmer Field Schools,[27] there has been some new thinking on what is generally termed "rural innovation systems."

A viable alternative to the T&V extension approach has yet to be found.

This approach has been supported by FAO and the International Service for National Agricultural Research (ISNAR) and is aimed at identifying constraints along the range of players in the commodity chain to develop a framework for prioritizing investments. The impact of this somewhat different angle on an old problem remains to be tested, then evaluated for impact on both growth and poverty.

Although creative ideas may be valuable, experience suggests there is some risk of grasping at fads. The Kenya agriculture sector review done for this study found that in both the adoption and the wholesale and abrupt abandonment of T&V by the Bank, there was excessive reaction to fads or pendulum swings, and insufficient revisiting of the core question of how the poor might be alternatively yet more efficiently served.

A newsletter of the nonprofit Sasakawa Africa Association (2005) noted that private contracting of agricultural advisory services has gained momentum in Africa partly because organizations such as the World Bank are championing this approach. The newsletter noted that key questions remain unanswered: "Will private contracting lead to improved conditions of employment for contracted extension staff? Will there be greater accountability in responding to farmers' needs and demands? Will cost recovery from farmers or from subsistence staple food crops be possible? Finally, will larger proportions of farmers be reached through contractual agreements on performance standards?"

Pluralistic extension approaches, though popular, face several challenges.

While pluralistic extension approaches have become very popular, their implementation faces several challenges. The transition from completely public-funded programs to alternative extension modalities with improved incentives also requires a significant investment of time—on the order of decades (Chapman and Tripp 2003). Uganda is in many respects at the forefront in Africa in developing a new demand-driven program in agricultural extension, and its experience illustrates some of the challenges (box 5.3).

And their sustainability and cost effectiveness have yet to be established.

The experience of other countries also demonstrates that it is not easy to implement extension approaches that are dependent on strengthening producer organizations and on contracting the services of private or semi-private service providers. The completion report for the Senegal Agricultural Services and Producers' Organizations Project (fiscal 1999) noted that although a semi-private agency for agricultural advisory services has been created to replace the former extension services, the agency was not completely accountable to producers. Moreover, activities that provided direct support to producer organizations were not given priority in the work of the agency. Some of the latest supervision findings from efforts such as the Kenya Agricultural Productivity Project (fiscal 2004) note issues that still need to be sorted out, including the need to develop the capacity of service providers, how to ensure transparency and fair competition in awarding contracts in weak institutional environments, and how service fees are to be determined, among others.

Maintaining the quality of Bank extension support, with multiple service providers, remains perhaps the greatest challenge. The appraisal document for the Zambia Agricultural Development Support Project (fiscal 2006) acknowledges, "In some instances, extension service provided by NGOs has reduced the control that the agribusiness companies have over the standard of service provided or the content of the technical advice and assistance being given. This has often resulted in inconsistent advice being given, causing confusion and having a negative impact on production. The situation with NGO or donor involvement in extension services is exacerbated when the project or funding ends and there is no sound exit strategy to ensure that service continues to be provided in a sustainable manner" (World Bank 2006m, p. 12).

Of critical importance to high-quality extension support is the training of extension service providers. IEG's assessment of the Tanzania Second Agricultural Extension Project (fiscal

Box 5.3: New Uganda Extension System Improves Efficiency But Faces Challenges

As the National Agricultural Advisory Services (NAADS) launched in 2001 expands, it is expected to replace Uganda's old extension system, which continues to function in the districts not covered by NAADS.[a] A midterm review of the program in late 2005 found NAADS to be more cost-effective than the earlier system.

Despite its apparent efficiency, the new system also faces several challenges, the report notes. These include inadequacy of service providers and resource constraints to implement NAADS effectively. Most of the funding support for NAADS currently comes from donors, but local governments and farmers are expected to take on increasing responsibility over time. However, the many stakeholders in the system have an imperfect understanding of the NAADS principles and it is unclear whether the local governments will have the resources to take responsibility for supporting NAADS.

The midterm report recognizes that the program "will certainly face challenges as it scales up to nationwide coverage." Drawing on their work in Uganda, Ellis and Bahigwa (2003) note that "while there has been a move away from top-down prescriptive support to sectors or subsectors, there is now far too great a reliance on an idealized concept of participatory processes in communities to enforce good governance on the part of local councils and effective service delivery by public agents at the local level."

Bahigwa and others (2005) are also concerned about the ability of NAADS to reduce the disadvantages of the poor in comparison with the nonpoor. Finally, Whyte and Kyaddondo (2006) found that despite successive initiatives, neither access to extension services nor technology adoption has reached 1970 levels.

a. NAADS is expected to facilitate formation of local farmer groups and farmer forums at the subcounty, district, and national levels. The farmer groups are expected to articulate their needs and fill them through purchases from private sector providers. The services are to be paid by the public sector through the decentralized local government institutions.

1997) notes that, at present, the strategy of "pluralism" appears to have an unspoken subscript that suggests that the approach will push private and NGO-supported extension and farmer-funded extension as far as it can go.

However, the shift to the private sector brings additional problems. Private and NGO-based extension services currently rely on buying and supplementing public extension by paying salary supplements and travel. If public extension did slowly die, NGOs and the private sector would need alternative, more costly approaches to access the same skills. In effect, they are free-riding on the underutilized skills, training, and salaries of the public extension service. Although this is efficient in the short term, it may not be sustainable in the longer term. Contracting out extension makes it possible to take advantage of all the experience in the field, but does not eliminate government's role. In addition to funding, government ensures quality assurance, oversight, and provision of information to contracted service providers (Muyanga and Jayne 2006). The need to ensure an adequate connection with research is also critical.

Training of extension service providers is critically important.

Effective M&E of Bank-supported projects will be necessary to help determine whether demand-driven and partnership approaches will be able to meet the needs of poor subsistence farmers. Private extension generally is skewed toward well-endowed regions and high-value crops, while remote areas and poor producers, particularly those producing low-value crops and little marketable surplus, are poorly served (Muyanga and Jayne 2006). The Kenya agriculture sector review undertaken for this study noted that extension in Kenya needs a realistic strategy and a clear role for the public element quite soon, otherwise it will wither and it will not be possible to bring it back.

Effective M&E will be needed to assess the efficacy of demand-driven approaches for poor farmers.

Post-T&V, it is unclear what a pluralistic approach to extension will mean for the poor. It is also unclear whether subsistence farmers (a large majority of whom are women) will be able to pay for the service provided, at least in the near future.[28] It is also difficult to tell whether it will be possible for them to organize effectively to create "demand" for extension services that will improve productivity of cassava, sorghum, millet, and other food crops.

The Bank and policy makers need to compare the cost effectiveness and appropriateness of various public and private extension options, including radio and television, for handling different short- and long-run opportunities and challenges for food and cash crop production.[29] A recent compilation of case studies on extension by the Bank's ARD Department (World Bank 2004b) also highlights the need to develop a better understanding of diverse approaches before reforms are undertaken.[30] There is a need to exercise a measure of caution in scaling up the demand-driven and partnership approaches before donors and borrowers can be reasonably sure that the returns will be commensurate with the costs and that the new approaches will not have to be "rejected" in the future. This process of comparing cost effectiveness and appropriateness does not have to be time consuming, but can be undertaken fairly quickly with critical borrower input.

Land Reform

Recent World Bank analytical work on land policy issues has contributed to the understanding of property rights regimes and their importance for agricultural development. Moreover, anecdotal evidence suggests that Bank policy advice has helped put land issues on the political agenda in many countries. For example, the agriculture sector review for Mali done for this study found that the government has rewritten the land tenure law to provide better land security and improve the likelihood of private investments in the land, and that this was undertaken in part because of the Bank. In most countries, though, the Bank has found it very difficult to provide lending support for land reform because it is a politically, socially, and culturally sensitive area.

Bank analytical work on land policy has contributed to better understanding of property rights regimes.

IEG was able to build a list of projects that dealt with land reform or land policy issues over the period 1991–2006 by combining research work done by ARD and the Land Policy Thematic Group. During 1999–2006 there were only four free-standing "land" projects: a Rural Land Management Project in Côte d'Ivoire (fiscal 1997), which has recently closed but for which there is no completion report yet; a Land Reform Support Project in Zimbabwe (fiscal 2000), which did not become effective; and two active projects, Ghana Land Administration (fiscal 2004) and Malawi Community-Based Rural Land Development (fiscal 2004). In all other cases, support for land reform is part of a wider environment or agriculture intervention.

There have been only four free-standing land projects.

Some PRSCs, such as the PRSC2 in Tanzania (fiscal 2005), have also attempted to develop a strategic plan for implementation of land reforms. In addition, a few emergency response interventions, such as the Eritrea Emergency Demobilization and Integration (fiscal 2001), have attempted to increase access to land for disabled soldiers, but it is too early to say how successful these interventions have been.

Among the findings of IEG assessments and Project Completion Reports is that land reforms are important for ensuring broad-based growth. IEG's assessment of the Zimbabwe Second Structural Adjustment Project (fiscal 1993), in particular, noted that agricultural marketing reforms alone could not ensure such growth. The skewed distribution of land needed to be resolved because most of the benefits of the marketing reforms went to the few thousand commercial farms that were able to respond quickly to them (IEG 2003d).[31] The Bank appears to have realized this long before the project was assessed. Immediately following the marketing reforms project, the Bank attempted to pilot an approach to land reform. However, implementation of that intervention was not easy (box 5.4).

The implementation of land reform interventions in other countries has also been complicated by socio-political factors. In the Côte d'Ivoire Rural Land Management Project the Bank provided support for titling of customary rights. However, it was not easy to document all "secondary" rights of the groups within the community. As a result, the project merely achieved a simplification of rights. This tended to strengthen the position of the individual landholder at the expense of the other right holders (van den Brink and others 2005). In another example, the Malawi Community-Based Rural Land Development Project (fiscal 2004) sought to increase the incomes of about 15,000 poor rural families by implementing a decentralized community-based and voluntary approach to land reform in southern Malawi. Progress toward the development objective was slow because of challenges in land acquisition and delay in surveys of farms to be acquired, among other things.

Bank project activities have generally shown inadequate appreciation of the time that is required to build consensus around sensitive issues such as land reform. The Lesotho Agriculture Policy and Capacity Building Project (fiscal 1998) had a component for facilitating the development of a new land policy and legislation compatible with sustainable land use systems. While the government made significant progress with respect to land policy, the new legislation was not enacted by project close. The project design had not accounted for the time-consuming stakeholder consultations required to reach consensus on land legislation.

In Ghana, the objective of the Land Administration Project (fiscal 2004) was to develop a sustainable and well-functioning land administration system that is fair, efficient, decentralized, and enhances land tenure security. Supervision missions have noted that the objective was ambitious for a five-year project, and at best the project could be a first phase that laid the foundation for accelerating reforms in the sector.

One lesson is that land reforms are important to broad-based growth.

Price and Marketing Reform

Reforming output and input prices and markets to improve the incentives for growth of agriculture has been a major area of Bank intervention in Africa. While a significant part of this reform was attempted in the late 1980s and 1990s through policy advice and structural and sectoral adjustment credits (now called development policy lending), sector projects have also been important. The adjustment reforms were meant to improve the incentives for farmers to increase production by reducing domestic market distortions and by encouraging private traders to replace the inefficient state trading companies (box 5.5). Since 1980 more than 30 countries have undertaken agricultural policy reforms as part of the broader adjustment agenda (Jayne and Jones 1997).

The Bank has generally underestimated the time required to effect reform around such a sensitive issue.

More than 30 countries have undertaken agricultural policy reforms since 1980.

Ex-post analysis, based primarily on the findings of Project Performance Assessment Reports, the portfolio review, country agriculture sector

Box 5.4: Zimbabwe Pilot for Land Reform Fails to Take Off

The Land Reform Support Project (fiscal 2000) in Zimbabwe was designed to pilot market-assisted land reform approaches. The project would have introduced a number of innovations for increasing direct participation of the ultimate beneficiaries and Rural District Councils in the planning and implementation of resettlement schemes.

Given the importance of the land issue in the country, the project's effectiveness date was extended four times to allow government to meet six (operational) conditions. However, these were not met and the credit was allowed to lapse.

The completion note for the project noted that following the amendment to the constitution and the Land Acquisition Act in mid-2000, the government's land reform strategy moved away from land acquisition at market value and the piloting of community-driven models to an approach based on compulsory acquisition at below-market value. The government lost the political will to go through with the agreed approach as the political situation in the country changed with the emergence of a strong opposition party. These developments completely undermined the project concept.

Box 5.5: Agricultural Market Reform in Africa: The Expectations

Structural adjustment began as a way to reform overspending parastatals, but it evolved to achieve other ends as countries' current account deficits increased. As the deflationary effects of higher import prices became clear, removing other price distortions (subsidies and taxes) and improving the regulatory environment for private entrepreneurs also became important. These measures were to improve the efficiency of resource allocation by having price signals accurately reflect their real values to society and by enabling private entrepreneurs to compete with and even replace parastatals.

The agriculture sector was important in the reform agenda for two reasons. First, it represented a substantial component of domestic production in most African countries, and supply response in the sector was a crucial determinant of the economy's response to changing incentives. Second, most economists and policy makers were convinced that trade and sector policies had been discriminating against the agricultural sector. Redressing this bias became a priority of the structural reform agenda. A healthy pattern of structural adjustment, based on exports and income expansion rather than on imports and demand contraction, was expected to stimulate strong agriculture sector performance. In most African countries, Bank-supported adjustment investment projects sought to phase out the provision of agricultural services better done by the private sector and support revision of regulations to provide an enabling environment for private sector investment in the agricultural sector. Broadly, the reforms were meant to:

1. Liberalize input-output prices by reducing or eliminating subsidies on agricultural inputs, realigning domestic crop prices with world prices, eliminating pan-seasonal and pan-territorial pricing, and reducing exchange rate overvaluation.
2. Remove regulatory controls in input and output markets, lifting restrictions on the internal movement of food crops and relaxing quantitative controls such as delivery quotas and licensing arrangements.
3. Restructure public enterprises and withdraw marketing boards from pricing and marketing activities and narrow their role to more supportive activities.

The *expected* long-run outcomes were:

4. Incentives for farmers are improved by increasing product prices and decreasing input costs, principally by encouraging private traders to substitute for the state trading companies.
5. Private investment is expanded.
6. Gains made in economic efficiency by eliminating price distortions and input price subsidies and the control of imports.
7. Trade balances are improved by stimulating exports and reducing imports.
8. Agricultural production and incomes for farmers are improved; better food security.

Sources: Sanders and others 1996; Mellor 1998; Kherallah and others 2002; study research.

reviews, and the evidence in the literature, finds that reforms have been pursued to varying degrees in different countries and points to both positive and negative influences flowing from the reform process. There were variable results across countries and crops. Although difficult to clearly categorize, there was comparatively more success achieved on some aspects than others. (appendix J summarizes reforms and achievements from Bank credits). In the literature there is consensus (Eicher 1999; Mkandawire and Soludo 1999 as referenced in Kherallah and others 2002; IFPRI 2000) that the reform program fell short of achieving its expected outcome.

The reforms have been pursued to varying degrees and had both positive and negative influences.

The reform process in Tanzania and several other African countries generally improved the macroeconomic environment and provided greater fiscal discipline through rationalization of the role of the public sector and promotion of a market-based exchange rate. According to IEG's 2003 *Annual Review of Development Effectiveness* (ARDE; IEG 2004b), policies in Africa have, on average, improved modestly, and those improvements have held. Analysis of country policy and institutional assessment (CPIA) data shows that overall CPIA ratings have improved for Africa since the late 1990s, but they remain below those of other Regions. IEG's project assessment for the Tanzania Agriculture Sector Management Project (fiscal 1994) concludes that in the broader institutional development sense, the "rules of the game" in Tanzania have changed substantially in a positive direction over the past decade, and the Bank project can claim to have contributed to this change.

The reforms also led to the withdrawal of marketing boards from pricing and marketing in several countries, relaxation of quantitative controls, and removal of regulatory controls in input and output markets. These changes considerably improved the incentives for production of some traditional export crops such as cotton. Growers of these crops in several countries are able to receive a greater share of the world price for the products (see appendix I for the story of cotton sector reform). The few studies available, some by the World Bank (Baffes 2005), generally confirm the positive change in marketing, particularly in cotton.

Overall, the picture was variable across countries and crops. For example, coffee production is reported to have increased in Uganda after the liberalization, while in Cameroon the policy reforms had a negative impact on the cocoa and coffee sectors (box 5.6). IEG's assessment of the Uganda Agricultural Sector Adjustment Credit (fiscal 1991) found that the project supported the very successful shift from the Coffee Board marketing monopoly to licensed private coffee traders. Following the change in marketing, coffee farmers, by the end of the project, were receiving 65 percent of the export price, compared with 30 percent before. In countries such as Mozambique, the story of cashews is much more complicated, as discussed in chapter 4.

The reform process also gave a boost to exports of nontraditional crops such as flowers from Kenya and mangoes from Mali. Today these crops represent a small but growing share of agricultural value added in several countries. The private sector has been playing an important role in this area. As with the rest of the agriculture sector, however, continued growth in nontraditional exports is challenged by weak institutions, poor transportation, and high input prices. Competition from countries outside of Africa is also a factor. Addressing increasingly stringent sanitary and phyto-sanitary standards in global markets is an even bigger challenge for Africa. There is a growing awareness of the need for supply chain coherence and efficiency in export marketing. While the Bank has been helping some countries in this area (Senegal, for example), there is still a long way to go.

Reforms led to the withdrawal of marketing boards from pricing and marketing in some countries.

Perhaps the biggest shortcoming is that the reform process had limited impact on food production. The average annual growth rate for agriculture value added was negative throughout the 1980s and 1990s (IFPRI 2000). In most reforming countries the private sector did not step in to fill the vacuum when the public sector withdrew. The portfolio review found that at least 30 percent of the ICRs reviewed raised this issue as a concern.

The reform process also boosted nontraditional crops, but had only limited impact on food production.

The private sector did not step in because of the prohibitive risks, high transaction costs, lack of access to information, and absence of contract and property right laws (IFPRI 2000). The project assessment of the Ethiopia National Fertilizer Sector Project (fiscal 1995) found that the project was not able to achieve its core objective of promoting a competitive fertilizer market because the private sector, already operating in a concentrated and government-dominated market, was squeezed out, and importing and distributing fertilizer became exclusively a government domain. "The inefficiency and misuse that prevailed during subsidy regimes prevalent in the pre-reform period have now been replaced by low profitability and high risk of fertilizer use" (IFDC 2006).

Input prices for the farmer rose dramatically. The value-cost ratios for a number of crops in several West African countries are reported to have declined since the 1980s, with most food crops having values of less than 2 in the mid-1990s (Heerink 2005).[32] Many otherwise viable technology options for Africa produced by past research remain underexploited because of high input and low output

The private sector did not step in because of high transaction costs, lack of access to information, and absence of contract and property laws.

Box 5.6: Negative Impacts of Policy Sequencing on Traditional Export Crop Sectors in Cameroon

The Bank played a significant role in the liberalization of cocoa and coffee marketing in Cameroon. In its first Structural Adjustment Loan to the country it initiated the breakup of the National Marketing Board (ONCPB) through loan conditionality. The reforms also called for devaluation, which finally occurred in January 1994. Unfortunately, because of the political economy of liberalization and the devaluation, the sequencing of these reforms was less than ideal for cocoa and coffee producers.

The removal of the 50 percent subsidy on fertilizers was the first policy reform implemented. Subsidies were gradually phased out from 1988 to 1992. This was followed in 1990 by a 40 percent cut in the official producer prices for coffee and cocoa by the ONCPB, which was unable to maintain stable producer prices in the face of the overvalued CFA franc and following the depletion of its reserves by a fiscally strapped government.

Producers responded by significantly curtailing resources allocated to cocoa and coffee agroforests, including fertilizers. With world coffee and cocoa prices at historic lows, state-administered panterritorial pricing was phased out in 1992 for coffee, but not until 1995 for cocoa. Price liberalization at a time of historically low world prices and an overvalued exchange rate resulted in farm gate prices that were less than half their nominal 1988 levels. At these prices, many farmers did not even harvest their coffee.

At the same time, the liberalization of fertilizer markets and the de facto liberalization of pesticide markets not only increased producer costs but also reduced availability because of the inadequately developed private sector. Fungicide control of cocoa blackpod disease fell dramatically, from over 30 million packets of fungicide distributed free by the state in the mid-1980s to less than 3 million purchased from private suppliers in 1993.

Finally, when devaluation came in 1994 and doubled nominal producer prices, the supply response was muted by a decline in the productive capacity of cocoa and coffee plantations, which had been, at best, minimally maintained under the policy regime from 1989 to 1994.

In sum, the unintentional effect of the structural adjustment conditionality was to seriously handicap Cameroon's smallholder export sector by a significant depreciation of farmers' tree stock. When prices in the mid- and late-1990s rose, farmers were unable to respond robustly.

Instead of asking where Cameroon's future comparative advantage was likely to lie in the late 1980s and 1990s, when commodity prices were at historic lows, and perhaps deciding that the economically important coffee and cocoa sectors of smallholder producers might have required support to help them adjust to a temporary shock in world commodity markets, the Bank pushed for its standard liberalization package. As a result, the country witnessed a serious extensification of its coffee and cocoa agroforestry systems.

Source: Essama-Nssah and Gockowski 2000.

prices (InterAcademy Council 2004). Analysts note that in promoting agricultural development, African governments have an important role to play in input-output market information systems, tax reforms, and regional cooperation where markets are too small to attract private investments. When these services are missing, the private sector cannot grow to its potential (Breman and Debrah 2003).

Most food in Africa is produced for home consumption by women farmers, who are not likely to be directly affected by the positive gains in the macroeconomic environment flowing out of the reforms. Farmers were, however, negatively affected by the rise in fertilizer prices. The rationalization and privatization of the work of the cotton and other parastatals further worsened access to fertilizers. For example, the Mali agriculture sector review found that both (cotton and cereals) crop types received input, credit, and extension support from the cotton parastatal in its area of operation. However, the parastatal narrowed its range of activities to focus on core cotton operations in the late 1990s, and the support for cereal crops was discontinued. As a result, fertilizer use for maize and other cereals declined sharply between 1999 and 2000.

Farmers were negatively affected by high fertilizer prices.

The large imports of cereals undertaken by several countries to meet the needs of the domestic market have led to a serious drain of foreign reserves in many countries in Africa. As a result, the expected improvement in trade balance noted in box 5.5 did not materialize.

High input prices have also adversely affected export crops. Before the reforms, particularly in West Africa, the parastatals dealing with crops such as cotton used to meet the input and credit needs of the farmers and assured them a secure market for their outputs. With privatization, producers of export crops in some countries are now faced with the same constraint as food crop producers with regard to access to inputs and credits. For example, the Senegal Country Assistance Evaluation (IEG 2006f) found that the private sector failed to engage in input supply, commercialization, or marketing following liquidation of the groundnut parastatal that had been active in distribution of seeds and fertilizers and the collection of groundnuts. This reduced farmers' access to critical inputs.

While in some countries organizations of producers have come up with Bank support to address this issue on behalf of their members, this has not happened across crops or in all countries. One reason has been the time it takes to build efficient producer organizations. Also, given the diversified cropping patterns in Africa, it has not always been possible for farmers to form single-commodity associations.

Why did results fall short of expectations? Because of inadequate background analytical work, weak political support, and insufficient appreciation of the system's incentives.

With regard to inadequate background analytical work, Tshibaka (2003, pp. 275–76), commenting on the privatization process supported under Bank projects, notes that "Little attempt was made to identify functions that are best performed by government agencies and those that are best handled by the private sector or to assess the private sector base in each country concerned. The failure to examine these and other related key questions has made it difficult for the designers of the structural adjustment reforms to propose appropriate policy measures and actions that could help strengthen and foster the development of the private sector in order to enable it to effectively handle various functions that were previously carried out by parastatals in the economy."

High input prices have adversely affected export crops.

Tshibaka's finding is supported by the evidence from the recent project assessment of the Agriculture Sector Management Project in Tanzania (fiscal 1994). The assessment notes that "the issue was not merely what activities could be best carried out by the central ministry(s), it was also what activities could be best carried out by the private sector, by partnerships, or by more independent commodity organizations, given the capacity of these alternative service mechanisms at the time" (World Bank 2007d, p. 8). While this emerges as a fundamental weakness in design in the Tanzania project, it was symptom of a wider problem with design of most similar projects.

IEG's 1998 Kenya CAE (IEG 2000b) also noted the failure of the Bank and borrower to focus sufficiently on the capacity of the private sector to pick up the roles left by divestiture. For example, the removal of the National Cereal and Produce Board (NCPB) monopoly, something that had been asked for since 1980, was not accompanied by enough analysis of what would happen afterward, given the poorly developed trader and storage network. While some companies did invest for a short period, the continued threat of NCPB intervention has kept them out of new investment for the past 10 years.

Results fell short because of inadequate analysis, weak political support, and insufficient appreciation of system incentives.

The Senegal Country Assistance Evaluation (IEG 2006f) also notes, "A major issue that delayed the liberalization of the groundnut sector has been whether the reforms could have adverse distributional consequences for poor farmers. The Bank should have undertaken analytical work on these issues sooner, given the importance of this sector to rural livelihoods" (p. 25).

Further, achievement of the full benefits of the process required active government and donor support to develop and integrate markets, not simply "liberalize" them.[33] This meant attention to the development of infrastructure to ensure coordinated and sustainable systems of input delivery, farm finance, and reliable output markets, not simply trusting the market to take over.

As already seen in chapter 4, the negative impact that weak political support and capacity in the borrower can have on the success of the reform process was not well appreciated. The weak political will among several governments led to partial adoption of reforms and delayed implementation, and even reversals in several cases.

At the sector level, policy makers saw incentives in terms of changing prices, whereas individual farmers were motivated by considerations of income, of which price and costs are a part (Donovan and Casey 1998). A large number of farmers whose product never enters the market did not benefit from improved output prices, but were adversely affected by input prices.

Chapter 6

Peanuts harvested in Mali. Photo by Ray Witlin, courtesy of World Bank Photo Library.

Findings and Recommendations

Key Findings

Agricultural development in Africa is a complex technical, economic, social, and political challenge that has to be overcome if the Region is to reduce extreme poverty and hunger—to meet the first Millennium Development Goal.

- Agricultural land in Africa falls into many agro-ecological zones and is generally characterized by poor soils, highly variable rainfall, and frequent droughts. Farmer access to irrigation and transport infrastructure is limited, as is their access to credit, improved seeds, and fertilizers. Extension support for improved soil and water management practices is weak. The majority of farmers are smallholders with 0.5 to 2.0 hectares of land and rely on diversified coping strategies that involve planting several crops with different maturity periods and keeping livestock. Total agricultural output in Africa consists primarily of food crops; agricultural export crops account for less than 10 percent of total production.

- Increases in agricultural production in the Region have mainly come from area expansion rather than yield increases and have not kept pace with population growth. In recent years, however, expansion too has stagnated, indicating that land frontiers may have been reached, at least in some countries. As land becomes scarce, issues of ownership and property rights become more important. Land ownership in most countries in Africa is determined by socio-cultural and political factors. Social factors also determine the division of agricultural labor between men and women.

- Political commitment to develop the agriculture sector has generally been low. Governments have used agriculture more as a source of resources for growth and have not invested adequately in its development. Government capacity is weak, and exchange rate and market distortions and poor incentives have limited private sector development and have also kept farmers from taking risks and intensifying agricultural production.

Given the diverse constraints to agricultural development in Africa, the strategy for the development of the sector needs to be multifaceted, with coordinated interventions across a range of activities.

- Farmers have to be convinced that it will be to their advantage to take on the newer technologies before they will undertake intensive agriculture to improve productivity and cultivate new crop varieties that are riskier and depend on the availability of fertilizers and water. To do this, a number of factors need to come together at the same time, or at least appear in an optimal sequence: improved seeds, water, credit, and access to markets; good extension advice; and adequate returns through remunerative prices for inputs and outputs.

- Public-private and donor partnerships need to be developed/strengthened, with actors contributing in areas of comparative advantage.

The Bank has had limited success in helping address the challenges of agricultural development in Africa.

- The institution's strategy for the development of the agriculture sector has been part of its rural strategy, and over time the importance of agriculture in the Bank's rural strategy has declined. While the broader rural focus by the Bank from the mid-1980s was justified, an unintended result was that it led to less focused attention on the need for various activities that are critical to agricultural development in rural space to come together at the same time or to take place in some optimal sequence. Both arising from and contributing to this, technical skills to support agricultural development adequately have also declined over time.

- The Bank's diagnosis of a country's development status and priorities is carried out primarily through analytical work. Until very recently this has been limited and not readily available. Nor have the findings from analytical work strategically informed Bank client policy dialogue and lending program design.

- The Bank's lending support has shown little recognition of the need to develop agriculture in a multifaceted way, but has been "sprinkled" across agricultural activities such as research, extension, credit, seeds, and policy reforms, with little apparent recognition of the synergies among them. Partly because it has not taken a multifaceted approach to agricultural development, the institution has not been able to take a long-term strategic approach to drought and food security. As a result, though there have been areas of comparatively greater success—research, for example—results have been limited because of weak linkage with extension and limited availability of such complementary inputs as fertilizers and water.

- The Bank's data systems and support for M&E have been insufficient to adequately inform the institution's effort to develop agriculture in Africa across a broad front. Current data systems do not allow the institution to track in enough detail how much is being provided for development of specific activities related to agriculture. M&E at the project level has been of limited value in answering fundamental questions about outcome, impact, and efficiency, such as who benefited, which crops received support and how, what has been the comparative cost effectiveness, and what is the appropriate attribution of the gains.

Recommendations

The Bank has an opportunity to contribute in a major way to development of African agriculture because it is one of the largest sources of development finance, and no other international donor has the Bank's ability to provide policy advice to governments. To effectively support the implementation of the Africa Action Plan and its appropriate focus on agricultural development as a key priority, IEG recommends that the Bank:

1. **Focus attention to achieve improvements in agricultural productivity:**
   - Establish realistic goals for expansion of irrigation and recognize the need to increase productivity of rain-fed agriculture through improvements in land quality, as well as water and drought management.
   - Help design efficient mechanisms, including public-private partnerships, to provide farmers with critical inputs, including fertilizers, water, credit, and seeds.
   - Support the development of marketing and transport infrastructure.

2. **Improve its work on agriculture:**
   - Increase the quantity and quality of analytical work on agriculture and ensure that policy advice and lending are grounded in its findings.
   - Support public expenditure analyses to assess resource availability for agriculture and to help set Bank priorities.
   - Rebuild its technical skills, based on a comprehensive assessment of current gaps.

3. **Establish benchmarks for measuring progress:**
   - Improve data systems to better track activities supported by the Bank.
   - Strengthen M&E to report on project activities in various agro-ecological zones and for different crops and farmer categories, including women.
   - Develop a system to coordinate agriculture activities in a country with road access, market proximity, and soil conditions.

Woman waters single plant, Ethiopia. Photo by Ray Witlin, courtesy of World Bank Photo Library.

Appendixes

The evaluation used four main instruments: a review of the Bank's lending and nonlending activities; country-level agriculture sector reviews; a review of relevant Bank and non-Bank literature; and a survey of Bank staff. In addition, IEG conducted 13 assessments of agricultural projects in various African countries during fiscal 2007.

This appendix describes these instruments. Also included is a section on how IEG identified the Bank's strategic approach in Africa's agriculture sector.

Portfolio Review

The portfolio review was a desk study of projects in the Sub-Saharan Africa agriculture portfolio. The study team first identified all Africa agriculture projects, and then selected a sample for a detailed review of appraisal and completion documents.

Identifying the agriculture portfolio

The review covers the 15-year period from fiscal 1991 to 2006 and is restricted to projects funded by the IBRD and IDA. Using World Bank data, the study team identified the Africa agriculture portfolio using standard Operations Policy and Country Services (OPCS) sector codes, consistent with the methodology used by the Agriculture and Rural Development (ARD) Department to report on lending trends in the sector. The agriculture codes are grouped under two sectors: *Agriculture, Fishing, and Forestry* and *Industry and Trade.* The subsectors under the former are agriculture, extension, and research (AB); animal production (AJ); crops (AH); forestry (AT); irrigation and drainage (AI); and general agriculture, fishing, and forestry (AZ). Relevant subsectors under the industry and trade sector are agriculture marketing and trade (YA) and agro-industry (YB).

As noted by ARD, problems with the Bank's sector coding system may cause underreporting of lending to the agriculture sector. Investments for agriculture agency reform, land administration, and rural finance in particular may not be fully captured by sector codes.

In the Bank's database, sector codes are mutually exclusive, but thematic codes are not. Therefore, thematic codes have been used to identify projects for more detailed examination, but not for purposes of reporting on lending amounts.

Selection of sample for portfolio review

The database identified 262 agriculture projects approved for Africa during fiscal 1991–2006. However, the database does not recognize a supplemental project as an additional project—only the loan/credit amount is included. The logic is that since the parent project is already in the system, there is no need to count the supplemental separately. The IEG review therefore included supplemental projects when their parent project was approved before fiscal 1991 as a separate project. If the parent was approved during the study period (fiscal 1991–2006), the supplemental was not counted as a separate project, because this would have led to double counting. Hence, 10 additional supplemental projects were added to the identified universe, for a total of 272 projects.

A stratified random sample of 71 projects was selected from the universe of 272 projects for further review. The stratification used two criteria: the number of subsectors and country type. The sample comprises 54 investment projects and 17 adjustment projects. During the review of the sample, we discovered that two projects had been miscoded as agriculture. One, Benin Urban Rehabilitation and Management (fiscal 1992, P000097), was coded as AI, but the component related to cleaning of storm gutters was not agriculture-related irrigation and drainage. The other, Kenya El Niño Emergency Project (fiscal 1999, P056595) was coded as AI, but project components related to reconstruction of rural water supply (wells, culverts) were not for water for agriculture. These projects were replaced with the next two consecutive projects in the random number list: Uganda Agriculture Sector Management (fiscal 2002, P073604) and Eritrea Emergency Demobilization and Reintegration (fiscal 1996, P037582).

Other components of the portfolio review

IEG Implementation Completion Report (ICR) reviews: Extensive analysis of project performance was done for the 144 completed Africa agriculture projects using IEG ICR reviews. The analysis focused on lessons learned from agriculture projects, reasons for less than satisfactory Bank and borrower performance, and sustainability issues.

Monitoring and evaluation (M&E) analysis: In January 1996, clarifications from OPCS provided guidance to staff on preparing indicators. Accordingly, the 54 investment projects in the sample of 71 were examined for the extent to which the OPCS guidance had been internalized in project design and implementation and how the trend shifted after 1996. The indicators were categorized into three groups:

1. ***Output indicators***—mainly quantitative targets such as number of markets established, number of extension workers, number of smallholders, number of associations established, number of farmers/beneficiaries reached, number of loans, number of village banks established, reduction in fertilizer subsidies, and the like
2. ***Outcome indicators***—for instance, improved capacity of relevant ministry, improved research capacity, improved adoption of fertilizers, improved credit access, increased access to extension services, sale of parastatals, increased seedling production, and so on
3. ***Impact indicators***—such as increased productivity, increased land fertility, increased cultivated area, increased food security, improved trade balance, increased farmer income, and the like.

The review focused only on agriculture-related indicators. The actual share of agriculture varies considerably across projects, so we did not quantify the number of indicators included in the project documents, and the inclusion of even a single indicator is recorded in the analysis.

Human Resources data: Data for staff mapped to ARD in the Africa Region were obtained from the Human Resources (HR) Department. Staff was categorized as either economists and generalists or technical, based on their title.

Review of Quality Assurance Group (QAG) data: QAG Quality at Entry Assessment (QEA) and Quality of Supervision Assessment (QSA) reports were reviewed for all Africa agriculture projects in the portfolio that have been assessed by QAG. Thirty-seven projects were reviewed for QEA and 43 for QSA.

Country-Level Reviews

CAS/PRSP review

CAS review: To assess the evolution of the focus on agriculture and agriculture-related issues in the Bank's country strategies, two CASs (Country Assistance Strategies) were reviewed from each of the countries. The selection was made based on the availability of a CAS for a country from two periods, one from the 1990s and one from the 2000s. Because Sierra Leone only has a CAS during the latter period, the comparison could not be made for that country. Thirty CASs were reviewed for the remaining 15 countries, for a total of 31 CASs.

PRSP review: Sixteen African countries had completed Poverty Reduction Strategy Papers (PRSPs) as of July 2006. The selected documents were used to assess the borrowers' focus on agriculture and agriculture-related issues.

In-depth program review

The Bank's total lending program was reviewed in four countries where there has been significant Bank lending for agriculture. This was done to gain an in-depth understanding of the Bank's contribution to development of agriculture in those countries over time. For this analysis, two countries were selected in East Africa (Kenya and Malawi) and two in West Africa (Cameroon and Nigeria).

Project reviews

The review drew upon the findings of 13 project assessments (PPARs) carried out in fiscal 2007:

- Ethiopia: National Fertilizer Sector (ICR in fiscal 2003)
- Ethiopia: Seed System Development (ICR in fiscal 2003)
- Madagascar: Agricultural Extension Program Support (ICR in fiscal 2001)
- Madagascar: Irrigation Rehabilitation (PCR in fiscal 1995)
- Madagascar: Second Irrigation Rehabilitation Project (ICR in fiscal 2001)
- Malawi: Emergency Drought Recovery Project (ICR in fiscal 2005)
- Mali: Agricultural Trading and Processing Promotion Pilot (ICR in fiscal 2003)
- Mali: National Agricultural Research (ICR in fiscal 2002)
- Mali: Pilot Private Irrigation Promotion (ICR in fiscal 2004)
- Tanzania: Agricultural Research Project 2 (ICR in fiscal 2005)
- Tanzania: Agricultural Sector Management (ICR in fiscal 2002)
- Tanzania: National Extension Project 2 (ICR in fiscal 2004)
- Zambia: Emergency Drought Recovery Project (ICR in fiscal 2006).

These assessments provided the review with lessons of experience from the field as well as the perspectives of government officials and other stakeholders on the Bank's agriculture support in the countries involved.

Literature Review

A review of the relevant Bank and non-Bank literature was undertaken to provide a theoretical basis for understanding African agriculture and the Bank's role in its development. The literature review also provided a means for "testing" the findings emerging from the portfolio analysis and the country-level reviews.

A significant amount of research on issues relevant for agriculture and its development in Africa has been undertaken worldwide, including work by the World Bank. Given the diversity of conditions in Africa along with the varying potential for the growth of agriculture in the 47 countries in the Region, such triangulation of evidence is essential to answer the evaluation questions.

The review also built on relevant IEG evaluations, sector and thematic studies, and CASs, all of which are listed in the references at the end of this report.

Staff Survey

This instrument sought the views of relevant Bank staff on internal factors and incentives related to the Bank's assistance for agriculture in Africa. The staff survey was preceded by structured interviews of key staff in the Region and in ARD, which helped inform the design of the questionnaire. A total of 258 headquarters and country office staff and consultants were identified for the survey using the following criteria:

- ARD anchor staff and ARD-mapped staff in agriculture
- Water anchor staff and water-mapped staff in agriculture (excluding water and sanitation engineers, specialists, and financial analysts)
- Staff who are not primarily agriculture experts but have in some way contributed to agricultural development in Africa, as part of a team working on agriculture projects or on relevant transport, trade, or other sector investments; structural adjustment credits; sector work; or research.

The survey was e-mailed to the staff. The results of the survey were shared with management in the Region and in ARD. The response rate and survey results are reported in appendix F.

How the Bank's Strategic Approach in the Agriculture Sector Was Identified

The broad strategic goals the Bank has pursued in African agriculture over the period fiscal 1991–2006 were extracted from five rural strategy documents. The documents cover a wide range of issues and their treatment differs across documents. Table A.1, based on IEG's comparative analysis of the strategy documents, shows the set of critical constraints that defined the Bank's strategic approach to agricultural development in Africa during the period.

Table A.1: Identification of Key Constraints/ Priority Areas for Agricultural Development

Constraint/priority	A Strategy to Develop Agriculture in Africa, 1993	Vision to Action, 1997	Action to Impact: Africa Region Rural Strategy, 2002	Reaching the Rural Poor, 2003	Africa Action Plan, 2005
Need for price and market reform	X	X	X	X	
Research	X	X	X	X	X
Extension	X	X	X	X	
Natural resource management					
Soil degradation/conservation, soil fertility	X	X	X	X	X
Water management systems/ conservation	X	X	X	X	
Irrigation	X	X	X	X	X
Drought is covered but risk and vulnerability are seen as a broader issue	X	X	X	X	
Food security	X	X		X	
Agro-ecological diversity	X	X	X	X	
Transportation infrastructure	X	X	X	X	X
Credit	X	X	X	X	
Land policy/reform	X	X	X	X	

APPENDIX B: CATEGORIZATION OF COUNTRIES BY FACTOR ENDOWMENTS AND AGRICULTURE'S SHARE OF GDP

Table B.1: Cross-Country Typology for Sub-Saharan Africa

	Agriculture's share above average (34% GDP)		Agriculture's share below average (34% GDP)		
	Falling GDP per capita (1991–2001)	**Rising GDP per capita (1991–2001)**	**Falling GDP per capita (1991–2001)**	**Rising GDP per capita (1991–2001)**	**Middle-income countries (> US$1,000 per capita)**
More favorable agricultural conditions (top two-thirds of FAO country-level farming system assessment)					
Coastal country	The Gambia Togo	Benin Ghana Guinea-Bissau Tanzania	Côte d'Ivoire	Kenya Mozambique Senegal	Mauritius South Africa
Landlocked country		Burkina Faso Ethiopia Malawi Uganda		Lesotho Zimbabwe	Swaziland
Mineral-rich country	Cameroon Central African Rep. Congo, Dem. Rep. of Sierra Leone	Sudan	Angola Congo, Rep. of Zambia	Guinea Nigeria	Equatorial Guinea
Less-favorable agricultural conditions (lowest third of FAO country-level farming system assessment)					
	Comoros Burundi Niger	Mali Rwanda Chad	Madagascar	Mauritania	Cape Verde Botswana Gabon Namibia

Source: Diao and others 2006.

Note: Of the 47 African countries, the table does not include the following: Eritrea, Liberia, São Tomé and Principe, Seychelles, and Somalia.

Table B.2: Share of Agriculture and Agricultural GDP Growth Rates

Country	Agriculture's share of GDP (percent)		Average annual growth (percent)	
	1990	2004	1990–2000	2000–04
Better performers (average annual growth rate for 2000–04 greater than 5%)				
Angola[a]	17.9	8.5	–1.4	13.7
Mozambique[b]	34.1	21.2	4.8	8.9
Niger	35.3	0.0	3.0	6.4
Cameroon	24.0	40.0	5.5	6.0
Benin	36.1	32.1	5.8	5.7
Congo, Republic of	12.9	6.0	1.0	5.5
Nigeria	32.4	16.3	3.4	5.3
Burkina Faso	27.8	30.8	4.2	5.1
Gabon	7.3	8.1	–1.4	5.1
Mali	44.1	33.4	2.6	5.1
Ghana	44.8	35.3	3.4	5.0
Medium performers (average annual growth rate for 2000–04 greater than 2% and less than 5%)				
Tanzania	42.0	42.3	3.2	4.9
Rwanda	32.5	41.2	2.6	4.7
Guinea	23.4	24.3	4.6	4.5
Uganda	53.3	29.5	3.7	3.9
Guinea-Bissau	56.9	63.4	3.9	3.3
Central African Republic	43.8	57.0	3.8	3.0
Mauritius	11.0	5.4	–0.5	2.8
Togo	33.8	41.2	4.0	2.7
Poor performers (average annual growth rate for 2000–04 less than 2%)				
Burundi	51.1	36.1	–1.6	1.9
Kenya	25.3	23.9	1.9	1.9
Malawi	38.5	33.7	8.6	1.8
Botswana	4.5	2.3	–1.2	1.5
Madagascar	26.1	26.2	1.8	1.3
Zambia	18.2	18.8	4.2	1.3
Namibia	10.6	9.0	3.8	1.2
Ethiopia	50.7	42.2	1.9	0.9
Côte d'Ivoire	32.5	22.1	3.3	0.5
Senegal	19.9	17.0	2.9	0.0
Gambia, The	24.3	30.0	3.3	–0.2
Mauritania	26.6	17.0	4.4	–0.3
Swaziland	10.8	6.5	1.2	–0.3
South Africa	4.2	2.7	1.0	–0.4
Eritrea		12.6	1.5	–0.5

Country	Agriculture's share of GDP (percent) 1990	2004	Average annual growth (percent) 1990–2000	2000–04
Lesotho	19.6	15.5	2.0	−1.8
Zimbabwe	14.8	14.2	4.3	−9.0
Average annual growth rate not available				
Cape Verde	14.4	6.8	—	—
Chad	27.9	23.5	4.4	—
Comoros	39.4	36.2	—	—
Congo, Democratic Republic of	30.1	47.4	1.2	—
Equatorial Guinea	58.9	—	—	—
Liberia	54.4	54.9	—	—
São Tomé and Principe	27.6	18.5	—	—
Seychelles	4.8	2.6	—	—
Sierra Leone	44.0	43.2	−13.0	—
Somalia	62.7	—	—	—
Sudan	—	33.2	9.2	—

Source: 2006 World Development Indicators.

Note: For some countries high growth in agriculture is due to returns from activities in the forestry sector such as logging and growth in export crops.

a. The high growth rate is due to the process of rehabilitation and reactivations after conflict.

b. Agricultural growth has mainly been driven by the post-conflict resettlement of refugees in the rural areas and the resulting expansion in labor and land (World Bank 2006g).

Table B.3: Agricultural GDP Growth Rates (countries with average annual growth rate over 3%)

Country	Average annual growth (percent) 1990–2000	2000–04
Benin	5.8	5.7
Burkina Faso	4.2	5.1
Cameroon	5.5	6
Central African Republic	3.8	3
Ghana	3.4	5
Guinea	4.6	4.5
Guinea-Bissau	3.9	3.3
Mozambique[a]	4.8	8.9
Niger	3	6.4
Nigeria	3.4	5.3
Tanzania	3.2	4.9
Uganda	3.7	3.9

Source: 2006 World Development Indicators.

Note: For some countries high growth in agriculture is due to returns from activities in the forestry sector such as logging and growth in export crops.

a. Agricultural growth has mainly been driven by the post-conflict resettlement of refugees in the rural areas and resulting expansion in labor and land (World Bank 2006g).

Table B.4: Selected Agricultural Indicators for Africa, Asia, and Latin America

Indicator	Sub-Saharan Africa[a]	South Asia	Latin America
Irrigated area (% of cropland)			
1989–91	3.6	33	11.1
2001–03	3.6	39	11.4
Fertilizer consumption (100 gms per hectare of arable land)			
1989–91	142	745	602
2000–02	123	1,066	895
Agricultural machinery (tractors per 100 sq. km of arable land)			
1989–91	20	62	121
2001–03	13	130	123
Cereal yield (kilograms per hectare)[b]			
1993–95	1,034	2,128	2,493
2003–05	1,087	2,505	3,159
Food production index (1999–2001 = 100)			
1992–94	81.7	80.3	79.1
2002–04	105.9	103.5	110.4
Agricultural productivity[c]			
Agriculture value added per worker (2000$)			
1992–94	294	364	2,234
2002–04	341	401	2,812

Source: 2006 World Development Indicators.

a. Includes South Africa.

b. Cereals include wheat, rice, maize, barley, oats, rye, millet, sorghum, buckwheat, and mixed grains.

c. Calculations include cash crops and forest and fisheries.

APPENDIX C: AGRO-ECOLOGICAL DIVERSITY, PRODUCTION SYSTEMS, AND GROWTH RATES OF FOOD AND CASH CROPS

Table C.1: Extent of Major Climatic Zones and Agricultural Land Use in Africa

Climate zone or region	Total area, million hectares (percent)	Annual rainfall (mm)	Land use, farming systems, and main agricultural constraints
Desert	822.0 (29.1)	<100	Nomadic pastoralists and hunter/gatherers, camel, sheep, goats. Too dry and hot for agriculture.
Arid North Africa and areas in southern Africa	844.0 (17.1)	100–400	Nomadic pastoralists, sheep, goats, camel, and some cattle. Main crops are rice, wheat, barley, and sorghum. Production-based irrigation. High animal population, overgrazing, deforestation causing soil degradation. Frequent drought.
Semi-arid Southern Africa Sudano-Sahelian	233.0 (8.1)	400–600	Nomadic pastoralists. Millet/sorghum, cowpea, groundnut, cotton, some maize. Low potential for rain-fed agriculture and variable annual rains. Production mainly based peri-urban systems. Pervasive soil nutrient mining.
Dry subhumid Subhumid west Subhumid south	314.0 (11.0)	600–1,200	Zone of arable crop production – maize, sorghum, groundnut, cassava, sweet potato, cowpea, rice, tobacco, cotton, tea, soybeans, cocoa. Some animals – cattle, sheep, and goats. Declining yield, severe land degradation and soil nutrient mining. High degree of deforestation and use of marginal lands.
Moist subhumid Mountain east	584.0 (20.4)	1,200–1,500	Transition zone with cereals (maize) and root crops (cassava, yams), banana, pineapple, and sugar cane. Wheat, coffee in east African highlands. Livestock. High erosion potential and soil fertility limitations.
Humid Humid west Humid central Wetlands	409.0 (14.3)	>1,500	Tree crop zone – oil palm, rubber, cocoa, food crops, yams, cassava, banana, rice, pineapple, and forest resources. Severe disease infestations, which limited exploitation of crops and livestock. Low fertility of soils.

Source: Henao and Baanante 2006.

Table C.2: Production and Farming Systems of Sub-Saharan Africa

Farming system	Land area (% of Region)	Agricultural population (% of Region)	Principal livelihoods
Maize mixed	10	15	Maize, tobacco, cotton, cattle, goats, poultry, off-farm work
Cereal/root crop mixed	13	15	Maize, sorghum, millet, cassava, yams, legumes, cattle
Root crop	11	11	Yams, cassava, legumes, off-farm income
Agro-pastoral millet/sorghum	8	9	Sorghum, pearl millet, pulses, sesame, cattle, sheep, goats, poultry, off-farm work
Highland perennial	1	8	Banana, plantain, enset, coffee, cassava, sweet potato, beans, cereals, livestock, poultry, off-farm work
Forest based	11	7	Cassava, maize, beans, cocoyams
Highland temperate mixed	2	7	Wheat, barley, teff, peas, lentils, broadbeans, rape, potatoes, sheep, goats, cattle, poultry, off-farm work
Pastoral	14	7	Cattle, camels, sheep, goats, remittances
Tree crop	3	6	Cocoa, coffee, oil palm, rubber, yams, maize, off-farm work
Commercial – largeholder and smallholder	5	4	Maize, pulses, sunflower, cattle, sheep, goats, remittances
Coastal artisanal fishing	2	3	Marine fish, coconuts, cashew, banana, yams, fruit, goats, poultry, off-farm work
Irrigated	1	2	Rice, cotton, vegetables, rain-fed crops, cattle, poultry
Rice/tree crop	1	2	Rice, banana, coffee, maize, cassava, legumes, livestock, off-farm work
Sparse agriculture (arid)	18	1	Irrigated maize, vegetables, date palms, cattle, off-farm work
Urban based	<1	3	Fruit, vegetables, dairy, cattle, goats, poultry, off-farm work

Source: Dixon and others 2001 in InterAcademy Council 2004.

Table C.3: Production of Food Crops in Agricultural Areas of Africa

	Growth per year 1995–2004		Area and yield 2002–04		
Region/crop	Area (%)	Yield (%)	Area (hectare)	Area (%)	Yield (ton/hectare)
Humid central					
Cassava	−0.6	−0.2	2,394,600	38.6	7.78
Maize	0.6	1.2	1,886,000	30.4	1.09
Pulses	1.3	1.3	605,900	9.8	0.81
Sorghum	−0.7	2.6	593,000	9.5	1.09
Humid and subhumid west					
Cassava	2.5	−0.6	5,433,700	12.4	9.51
Maize	−0.5	−0.8	7,404,000	17.0	1.13
Millet	1.6	−0.7	6,577,500	15.1	0.99
Pulses	3.2	−0.9	5,819,900	13.3	0.44
Rice	6.4	−2.7	6,513,000	14.9	1.14
Sorghum	1.4	0.1	7,927,000	18.2	1.03
Subhumid and mountain east					
Cassava	1.5	3.5	1,035,609	7.2	9.65
Maize	1.0	0.1	4,206,095	29.2	1.60
Pulses	2.1	0.0	3,729,910	25.9	0.70
Rice	0.9	0.7	1,351,290	9.4	2.29
Sorghum	2.7	−0.9	2,006,000	13.9	1.10
Wheat	1.3	0.1	1,178,200	8.2	1.41
Sudano-Sahelian					
Maize	6.7	2.4	1,191,483	3.7	1.41
Millet	0.5	3.1	12,192,458	37.9	0.53
Pulses	1.0	5.0	4,639,587	14.4	0.33
Rice	2.5	1.1	698,389	2.2	1.92
Sorghum	2.4	2.5	13,144,363	40.8	0.67
Subhumid and semi-arid southern					
Cassava	1.5	3.2	2,700,625	13.3	8.27
Maize	0.7	2.4	11,107,000	54.7	1.5
Millet	−1.0	0.8	1,055,021	5.2	0.54
Pulses	1.1	0.4	2,214,640	10.9	0.53
Sorghum	−0.8	1.0	1,619,342	8.0	0.96
Wheat	−3.7	4.4	1,033,550	5.1	2.29

Source: Henao and Baanante 2006.

Table C.4: Production of Cash Crops in Agricultural Areas of Africa

	Growth per year 1995–2004		Area and yield 2002–04		
Region/crop	**Area (%)**	**Yield (%)**	**Area (hectare)**	**Area (%)**	**Yield (ton/hectare)**
Humid central					
Coffee	−3.6	−0.6	360,000	15.4	0.32
Groundnuts	−1.7	3.0	841,000	36.1	0.88
Oil palm	0.5	0.6	329,000	14.1	7.83
Humid and subhumid west					
Groundnuts	4.5	0.8	3,900,500	35.0	0.96
Oil palm	1.0	−0.2	3,944,200	35.4	3.22
Seed cotton	4.3	−0.3	1,570,500	14.1	0.89
Subhumid and mountain east					
Bananas	1.2	−0.2	571,395	14.9	4.92
Barley	−4.5	2.9	767,000	20.0	1.19
Coffee	0.2	−0.8	955,670	25.0	0.61
Sudano-Sahelian					
Groundnuts	2.6	−0.5	3,943,550	68.8	0.73
Seed cotton	5.6	−0.4	1,610,461	28.1	1.06
Subhumid and semi-arid southern					
Groundnuts	3.4	0.6	1,193,137	31.1	0.54
Seed cotton	1.9	1.9	1,040,600	27.7	0.81
Sugar cane	1.2	0.3	474,000	12.3	70.78

Source: Henao and Baanante 2006.

The Portfolio

Table D.1: Details of Agriculture Lending to Africa, Fiscal 1991 to 2006

Total lending (all sectors) (US$ millions)	50,498
Investment lending (all sectors) (US$ millions)	34, 337
Lending to projects with agriculture components (US$ millions)	14,305
Lending to projects with agriculture components (as a percentage of total lending to Africa)	28
Lending for agriculture (US$ millions)	4,535
Lending for agriculture (as a percentage of total lending to projects with agriculture components)	32
Investment lending in agriculture (US$ millions) (Includes emergency recovery lending of US$247.22 million)	2,814
Investment lending in agriculture (including emergency) (as a percentage of total lending to Africa)	5.5
Investment lending in agriculture (as percentage of total investment lending to Africa)	8
Investment lending in agriculture (US$ millions) (Excludes emergency recovery lending of US$247.22 million)	2,567
Adjustment or development policy lending for agriculture (US$ millions)	1,721

Source: World Bank data.

Table D.2: Breakdown of Agriculture Lending by Region, Fiscal 1991 to 2006 (US$ millions)

Region	Total lending (all sectors)	Lending to projects with agriculture components	Lending for agriculture	Lending for agriculture as a percent of lending to projects with agriculture components
Sub-Saharan Africa	50,498	14,305	4,535	32
East Asia and Pacific	74,909	14,339	7,691	54
South Asia	50,764	12,818	5,808	45
Europe and Central Asia	63,380	11,120	4,446	40
Middle East and North Africa	19,713	4,815	2,731	57
Latin America and the Caribbean	86,138	11,156	4,601	41
Total	345,403	68,554	29,812	43

Source: World Bank data.

Figure D.1: Lending for Agriculture as Percentage of Total Lending to Projects with Agriculture Components, by Region

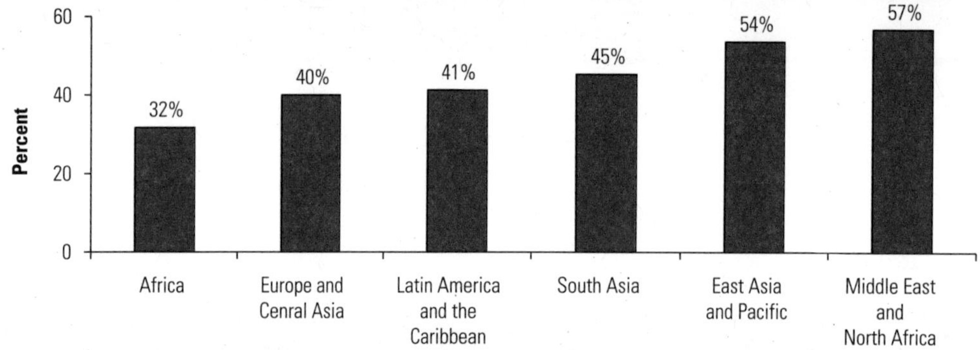

Source: World Bank data.

Figure D.2: Distribution of Bank-wide Agriculture Lending to Regions

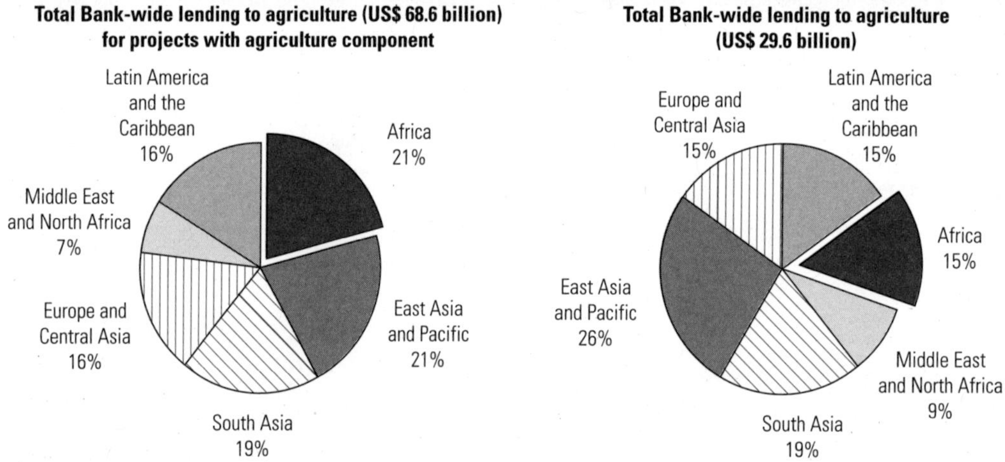

Source: World Bank data.

Figure D.3: Trends in IBRD/IDA Lending in Africa

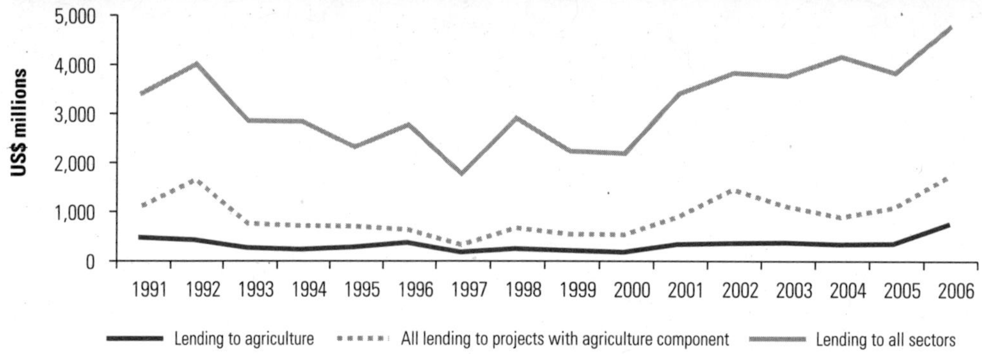

Source: World Bank data.

Table D.3: IBRD/IDA Lending to Africa, 1991 to 2006 (US$ millions)

1991	1992	1993	1994	1995	1996	1997	1998	1999	2000	2001	2002	2003	2004	2005	2006
Lending to agriculture															
419	374	221	178	234	323	127	196	160	123	286	308	318	287	295	685
All lending to projects with agriculture component															
1,088	1,625	730	683	675	609	301	634	512	502	862	1,415	1,068	855	1,054	1,695
Lending to all sectors															
3,379	3,971	2,815	2,808	2,284	2,740	1,730	2,871	2,205	2,159	3,370	3,793	3,737	4,116	3,792	4,727

Source: World Bank data.

Figure D.4: Lending to Agriculture by Subsector

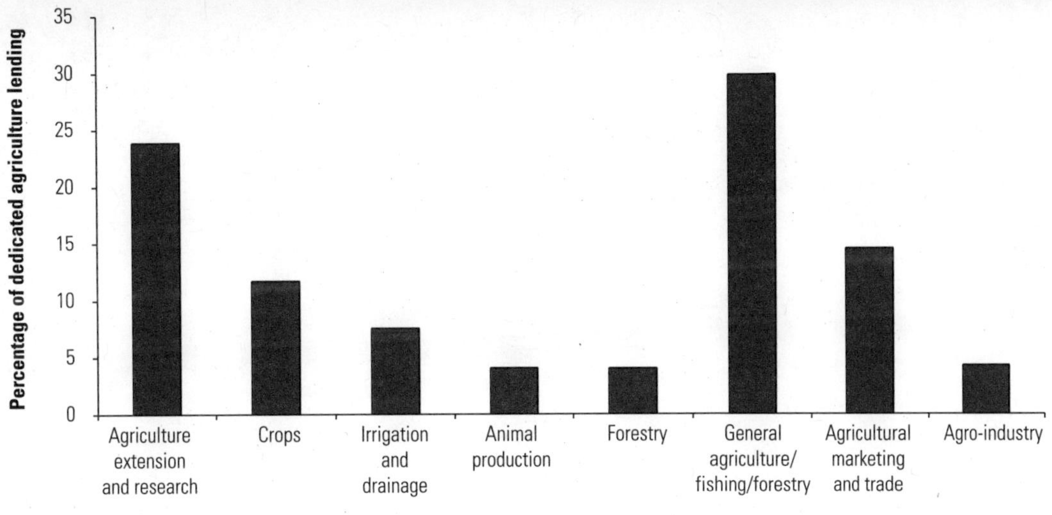

Source: World Bank data.

Table D.4 illustrates the point made in chapter 3 regarding the limitation of the World Bank's existing data systems. Information on Bank support at the country level is limited to the eight categories presented in the table. As noted in box 3.1, the current coding system is inadequate for tracking support to some critical activities that constrain agricultural development, such as seeds, credit, and land tenure.

Table D.4: Lending to Agriculture Subsectors, 1991–2006 (top 10 countries)

Fiscal year	Agriculture extension and research	Crops	Irrigation and drainage	Animal production	Forestry	General agriculture/ fishing/ forestry	Agri-cultural marketing and trade	Agro-industry	Overall result
Tanzania									
1991	1.0					1.6	5.0	6.0	13.5
1992	4.5					5.1	7.5	4.2	21.3
1993						0.2	0.2		0.5
1994	8.8								8.8
1997	5.6		14.5	2.5					22.6
1998	6.1								6.1
2000						13.3			13.3
2001						0.1			0.1
2002				1.3	5.3	4.1			10.7
2003	2.8	39.6				26.4	2.8		71.7
2005	0.3				0.3	32.1	30.0		62.7
2006	24.3		18.9			154.4	8.1		205.7
Total	53.5	39.6	33.4	3.8	5.6	237.4	53.6	10.1	436.9
Côte d'Ivoire									
1992						6.0			6.0
1994	12.2					3.0			15.2
1995							32.8		32.8
1996		39.4				43.1	109.1	10.5	202.0
1997						6.6			6.6
1999	42.5								42.5
2002		40.0					60.0		100.0
Total	54.7	79.4				58.6	201.8	10.5	405.1
Uganda									
1991	5.0	30.2		18.3			45.4		98.8
1992	3.6	0.2					0.3		4.0
1993	31.3								31.3
1994	4.6	16.0				15.7	6.4	0.6	43.2
1995		0.1				0.1	0.0		0.3
1996	1.3					14.0			15.2
1997	5.3			0.5		1.7			7.5
1999	10.4								10.4
2001	31.5						3.2		34.7
2002						3.9			3.9
2003	2.0			0.2		24.1			26.3
2004						22.5			22.5
2005						30.0		7.0	37.0
2006						16.2			16.2
Total	95.0	46.5		18.9		128.2	55.2	7.6	351.4

Fiscal year	Agriculture extension and research	Crops	Irrigation and drainage	Animal production	Forestry	General agriculture/ fishing/ forestry	Agri-cultural marketing and trade	Agro-industry	Overall result
Ethiopia									
1993		17.5							17.5
1994		0.0							0.0
1995	13.5	0.0					6.4		19.9
1996			12.0						12.0
1998	22.2								22.2
2001	1.8						40.9		42.7
2002						32.3	0.9		33.2
2003			2.8		6.6	45.0	12.0		66.5
2004						30.0			30.0
2006	62.5			0.5					63.1
Total	100.0	17.5	14.8	0.5	6.6	107.3	60.2		307.0
Ghana									
1991	37.5	0.3					0.2		38.0
1992	20.4	20.0				15.2	12.8	16.0	84.4
1993				16.2					16.2
1994						0.3	10.6		10.9
1995		1.3		3.0		1.0	0.8	1.0	7.0
1997						4.5			4.5
1998					4.3				4.3
1999						28.8			28.8
2000						0.2		1.6	1.8
2001	42.2					7.8			50.1
2002		11.0							11.0
2004						3.7			3.7
2005			12.0					12.0	24.0
Total	100.1	32.6	12.0	19.2	4.3	61.4	24.4	30.6	284.6
Mali									
1991	14.6	2.8		3.9	2.2				23.5
1992		4.1			4.1				8.2
1994	20.0								20.0
1995							5.6		5.6
1997			2.0						2.0
2000			52.9						52.9
2002	23.9	24.7		2.2					50.8
2004						3.0			3.0
2005			1.0					3.0	4.0
2006		11.1	2.5			29.4	26.0	2.5	71.5
Total	58.6	42.7	58.5	6.1	6.3	32.4	31.6	5.5	241.6

(Continues on the following page.)

Table D.4: Lending to Agriculture Subsectors, 1991–2006 (top 10 countries) (continued)

Fiscal year	Agriculture extension and research	Crops	Irrigation and drainage	Animal production	Forestry	General agriculture/ fishing/ forestry	Agri-cultural marketing and trade	Agro-industry	Overall result
Kenya									
1991	15.4	3.0		12.5	11.9	16.5	18.8		78.1
1993		9.5				6.0			15.5
1996						1.5	3.1		4.6
1997	29.5			0.5		2.0			32.0
1999			2.0						2.0
2001		23.0							23.0
2002		0.2							0.2
2003				18.0		24.0			42.0
2004	40.0								40.0
Total	84.9	35.7	2.0	31.0	11.9	50.1	21.8		237.4
Malawi									
1991	1.1			3.8					5.0
1992		1.4				1.0			2.4
1993	29.8		4.6						34.4
1996		17.0				24.5			41.5
1997		0.5				0.8			1.3
2001		6.7				0.4			7.0
2003						31.0			31.0
2004		12.5				24.0	12.5		49.0
2006	8.0	7.5	12.0			14.0	7.5		49.0
Total	38.9	45.6	16.6	3.8		95.7	20.0		220.6
Madagascar									
1991				13.5			0.1		13.5
1992							0.1		0.1
1993							0.1		0.1
1994						1.6			1.6
1995	25.2		15.1						40.3
1997			4.2			30.0			34.2
1998	0.3								0.3
2001	11.6		18.7			16.5		18.7	65.5
2003						15.0			15.0
2004	0.1					18.0			18.1
2005						16.3			16.3
Total	37.2		38.0	13.5		97.3	0.2	18.7	204.8
Burkina Faso									
1991					5.3	5.6		5.6	16.5
1992		4.8				2.2	8.1	7.0	22.1
1998	34.3		2.1	2.1					38.4

Fiscal year	Agriculture extension and research	Crops	Irrigation and drainage	Animal production	Forestry	General agriculture/ fishing/ forestry	Agri-cultural marketing and trade	Agro-industry	Overall result
1999		0.8	3.4						4.1
2001						10.0			10.0
2003						7.0			7.0
2004						6.5			6.5
2005						9.0		9.0	18.0
2006	22.4		21.8			9.0	9.9	5.9	69.1
Total	56.7	5.5	27.2	2.1	5.3	49.3	18.0	27.5	191.7
Nigeria									
1991	40.6								40.6
1992	37.0								37.0
2002	0.9							24.2	25.1
2004	10.0	12.0	30.0	4.2		14.0			70.2
2006	12.0			6.5					18.5
Total	100.4	12.0	30.0	10.7		14.0		24.2	191.3

Source: World Bank data.

Performance of the Portfolio

IEG Ratings

Table D.5: Agriculture versus Non-Agriculture Projects in Africa: Outcome and Sustainability Ratings

Lending type	Number of projects with outcome ratings	Outcome, percent satisfactory	Number of projects with sustainability rating	Sustainability, percent likely
Africa investment lending (50% or more to agriculture)	52	60	47	40
Africa investment lending (non-agriculture)	378	65	343	53

Source: World Bank data.

Table D.6: Agriculture in Africa versus Agriculture Projects in Other Bank Regions

Region(s)	Number of projects with outcome ratings	Outcome, percent satisfactory	Number of projects with sustainability rating	Sustainability, percent likely
Africa investment lending (50% or more to agriculture)	52	60	47	40
Other Regions, investment lending (50% or more to agriculture)	150	73	138	63

Source: World Bank data.

Table D.7: Non-Agriculture in Africa versus Non-Agriculture in Rest of the Bank

Region(s)	Number of projects with outcome ratings	Outcome, percent satisfactory	Number of projects with sustainability rating	Sustainability, percent likely
Africa investment lending (non-agriculture)	378	65	434	53
Other Regions, investment lending (non-agriculture)	1,103	79	1,028	77

Source: World Bank data.

Table D.8: Change in Performance of Agriculture over Time

	Number of closed projects	Number of projects with outcome ratings	Outcome, percent satisfactory	Number of projects with sustainability rating	Sustainability, percent likely
Africa investment lending (50% or more to agriculture)	60	52	60	47	40
1991–99	54	48	58	43	37
2000–06	6	4	75	4	75

Source: World Bank data.

APPENDIX E: LENDING TO AGRICULTURE FROM BILATERAL AND MULTILATERAL DONORS

Table E.1: Aid to Agriculture by DAC Countries and Multilaterals, 1981–2001

	Aid to agriculture (global) (% of donor total)			Aid to African agriculture (US$ million, 2001)			Aid to African agriculture (as a % of donor's total aid to Africa)		
Donor	**1981**	**1991**	**2001**	**1981**	**1991**	**2001**	**1981**	**1991**	**2001**
DAC countries	11	8	5	833	1,047	557	15	12	6
Multilaterals	33	22	8	1,089	640	440	32	14	7
Donors total	18	12	6	1,921	1,687	997	22	13	6

Source: OECD CRS database, as noted in Kane and Eicher 2004.

Table E.2: Aid to African Agriculture as a Percentage of Aid from all Donors to African Agriculture

	1990	**2000**	**2005**
Bilateral donors	61	58	52
Japan	11	11	6
United States	5	9	8
Multilateral donors	39	42	48
IDA	29	10	20
All donors	100	100	100

Source: OECD Creditor Reporting System.
Note: Excludes South Africa.

APPENDIX F: BANK STAFF SURVEY RESULTS

A survey was conducted to gather staff perceptions of institutional factors and incentives within the institution, as well as some general aspects of Bank support to agricultural development in Africa.

A total of 258 headquarter and country office staff and consultants were identified using the following criteria:

- Staff who are not primarily agriculture experts but have in some way contributed to agricultural development in Africa, as task managers or as part of teams working on agriculture projects or relevant transport, trade, or other sector investments, structural adjustment credits, sector work, or research
- ARD anchor staff and ARD-mapped staff in agriculture
- Water anchor staff and water-mapped staff in agriculture (excluding water and sanitation engineers, specialists, and financial analysts).

The survey was e-mailed to the staff and 56 responded (a response rate of 22 percent). Since it is in the nature of opinion surveys to have a response bias, it is difficult to ascertain whether those who responded are representative of the 258 staff to whom the survey was sent. Because of the limited number of responses and the likely response bias, the report has used the survey results only to substantiate findings from other information sources.

The survey response data are presented in table F.1. A brief analysis of the responses to the most pertinent questions follows the table. The survey also sought the views of Bank staff on some aspects of agricultural development through open-ended questions. The responses to these questions are presented after the analysis.

Table F.1: Bank Staff Survey: Response Rate

	Agree or strongly agree	Disagree or strongly disagree	Do not know
Strategic Approach to Agriculture in the Country Program			
1. The Ministry of Finance, which is the main counterpart for the Bank in the countries, recognizes the need for investment in agriculture development as a priority area for growth and poverty alleviation.	46.43	50.00	3.57
2. The current Country Assistance Strategies for countries in Africa generally reflect a strong focus on agriculture development.	26.79	69.64	3.57
3. The current Country Assistance Strategies are generally prepared in active consultation with agriculture staff in the Bank.	46.43	46.43	7.14
4. The Bank's policy dialogue bearing on rural development in the Africa Region adequately addresses technical issues in agriculture productivity (soil fertility, land management, land tenure, irrigation, improved seeds, etc.).	26.79	66.07	7.14
5. Sufficient and rigorous analytical work/sector work generally informs the design and implementation of agriculture projects in the Africa Region.	37.50	55.36	7.14
6. The strategic approach by the Bank towards focusing on rural development more broadly has diluted attention to technical issues in agriculture lending in the Africa Region.	64.28	25.01	10.71
7. The Bank's shift toward programmatic lending will sustain sufficient focus on technical issues in agriculture in the Africa Region.	19.65	69.64	10.71
Bank Support for Interventions in the Agriculture Sector			
1. It is much more difficult to show satisfactory results for agriculture sector projects in comparison to other sector interventions in the Africa Region.	57.14	35.72	7.14
2. Agriculture sector interventions are more complex and require longer-term support from donors than interventions in other sectors in the Africa Region.	78.57	17.86	3.57
3. The political economy in the countries in Africa is conducive for long-term support for development of agriculture.	44.64	53.57	1.79
4. Supervision and project preparation costs to the Bank for agriculture projects are significantly higher than projects in other sectors in the Africa Region.	62.50	19.64	17.86
5. Bank agriculture projects in Africa are able to respond adequately to the agro-ecological diversity and the needs of diverse production systems.	33.93	51.78	14.29

Bank's Strategic Approach to Agriculture

Only 26 percent of the respondents agreed that the current Country Assistance Strategies (CASs) for countries in Africa generally reflect a strong focus on agricultural development. There was no clear consensus among the respondents regarding whether the current CASs are generally prepared in active consultation with agriculture staff in the Bank.

More than 58 percent of the respondents disagreed that in the past decade the Bank has focused on priority issues for development of agriculture in Africa. Sixty-six percent of the respondents also disagreed that the Bank's policy dialogue bearing on rural development in the Africa Region adequately addresses technical issues in agricultural productivity (soil fertility, land management, land tenure, irrigation, improved seeds, and the like).

	Agree or strongly agree	Disagree or strongly disagree	Do not know
6. In the past decade the Bank has focused on priority issues for development of agriculture in Africa.	28.57	58.93	12.50
7. A focus on sustainability has been a significant element in project design for agriculture projects in Africa.	41.07	50.00	8.93
8. Community-driven development (CDD) approaches are effective in addressing critical sectoral issues in agriculture development in Africa.	42.85	46.43	10.72
9. The Bank's support for institution building in the agriculture sector in Africa, whether through T&V or CDD, has been carefully designed taking into account the reality on the ground and lessons of experience.	26.79	55.36	17.85
10. The Bank has a comparative advantage in the policy and institutional aspects to achieve satisfactory development outcomes for agriculture projects.	85.72	10.71	3.57
Bank Management Commitment to Agriculture Development			
1. The country directors in countries in the Africa Region sufficiently take into account the complex and multisectoral nature of agriculture activities in taking decisions on IDA allocations among sectors.	12.50	82.14	5.36
2. The current Bank matrix-management organizational structure adequately supports the needs of agriculture projects.	17.86	75.00	7.14
3. There is sufficient allocation of scarce IDA resources at the country level in the Africa Region for agriculture sector issues for optimal national development.	10.71	73.21	16.08
4. The Bank provides adequate resources overall (for lending and sector work) to support development of agriculture in Africa.	17.86	75.00	7.14
5. There is good coordination between donors working in the agriculture sector in countries in the Africa Region.	32.14	60.72	7.14
6. There is good coordination between staff working on agriculture and other sectors within the Bank in the Africa Region.	17.86	80.35	1.79
7. The Africa Region has an adequate level of technical staff skills (irrigation specialists, soil specialists etc.) to support implementation of agriculture projects.	17.86	67.86	14.28

Source: Staff survey.
Note: Based on 56 responses.

However, 85 percent of the respondents agreed that the Bank has a comparative advantage in the policy and institutional aspects to achieve satisfactory development outcomes for agriculture projects.

Complexity of the Sector

Seventy-nine percent of the respondents agreed that the agriculture sector interventions are more complex and require longer-term support from donors than interventions in other sectors in the Africa Region. Moreover, 57 percent of the respondents agreed that it is much more difficult to show satisfactory results for agriculture sector projects in comparison with other sector interventions in the Africa Region. More than 80 percent of the respondents disagreed that the country directors in the Africa Region sufficiently take into account the complex and multisectoral nature of agriculture activities in making decisions on IDA allocations among sectors.

High Cost of Agriculture Projects

Sixty-two percent of the respondents agreed that the supervision and project preparation costs to the Bank for agriculture projects are significantly higher than for projects in other sectors in the Africa Region. Seventy-five percent of the respondents disagreed that the Bank provides adequate resources overall (for lending and sector work) to support development of agriculture in Africa.

Bank's Internal Organization and Agricultural Development in Africa

Seventy-five percent of the respondents did not agree that the current Bank matrix management organizational structure adequately supports the needs of agriculture projects. More than 80 percent of the respondents disagreed with the statement that there is good coordination between staff working in agriculture and those working in other sectors in Africa.

Sixty-eight percent of the respondents disagreed that the Africa Region has an adequate level of technical staff skills (irrigation specialists, soil specialists, and so on) to support implementation of agriculture projects.

Responses to Open-Ended Questions

Q1. What do you consider to be the major constraint to agricultural development in Africa? In what areas has the Bank contributed to addressing these constraints?

The responses were categorized into the following groups:

Enabling Factors (those that "enable" agricultural development, such as roads that allow access to markets and credit that enables the farmer to buy seeds):

Lack of rural infrastructure (rural roads and irrigation) was identified by many respondents as a critical constraint for the development of agriculture in Africa. Lack of rural credit was next, followed by the lack of access to markets—both domestic and export. Other issues listed were inadequate extension or research and lack of private sector investment in agriculture.

In the view of one respondent, rural infrastructure issues are often ignored by agricultural staff in the Bank, who assume that they are being covered by colleagues in other sectors. Another respondent's view was that the Bank's portfolio does not address poor access to markets because it is not coordinated across sectors, and project locations rarely overlap, so synergies are not developed. Other reasons cited for neglect of attention to these issues within the Bank were:

- The Bank's emphasis on development policy lending and dialogue has been at the expense of action in critical productive sectors such as agriculture and infrastructure.
- Most country directors focus too much on PRSPs and PRSCs at the expense of investment projects.
- Most of the sectoral interventions outside the agriculture units (such as financial sector reforms, public sector reforms, energy, and transport) continue to have an urban bias, with insufficient attention to the development of agriculture.

Some respondents believed that infusion of funds through community-driven development operations is one option for development of small link roads, culverts, irrigation schemes, and watershed development.

Many respondents noted that the Bank has largely failed in addressing the credit needs of smallholders. In term credit and financial services, the Bank has consistently remained timid and very conservative. The financial sector family has been of little assistance in coming up with realistic and practical solutions to the problem of lack of or limited access to financial services to support real agriculture sector growth.

Incentive Factors (those that determine a farmer's incentive to produce):

Many respondents identified constraints such as a lack of incentives, a noncompetitive export sector, developed-country subsidies, an unfavorable business climate, and market distortions. Some respondents felt that the Bank's failure to address pricing issues at local, national, and international levels has adversely affected agricultural development in Africa. Insecurity of land tenure was also mentioned. According to some respondents, the Bank does not have any significant operations in Africa working on land tenure because of the political sensitivities surrounding the issue.

Physical Factors (availability of quality farmland, labor, and inputs, among others):

Among the physical constraints respondents identified were low agricultural productivity at the farm level, weak producer organizations, and human resource deterioration (such as HIV/AIDS, brain drain, low agriculture education and training investments, and so on).

A few respondents mentioned that the Bank portfolio is still too focused on the elements that made the Green Revolution work in Asia. They noted that this will work only in certain agro-ecological zones and political/institutional environments.

Natural Factors (weather and disease related):

Post-harvest losses, plant and animal diseases, and weather shocks were the three natural factors listed by some of the respondents who believe that the Bank needs to develop better strategies to help farmers cope with weather shocks.

Institutional Factors (government capacity):

A majority of the respondents noted institutional constraints: poor governance and weak institutional capacity, especially in the Ministry of Agriculture. Other constraints were weak agricultural policy frameworks and lack of sustained strategic priorities.

Respondents said that the Bank has not adequately addressed some of the major institutional constraints. They attribute this to: inadequate or insufficient analytical work, lack of assessment of past priorities, and unwillingness on the part of Africa's senior management to address deep-seated issues of political economy. Some respondents acknowledge that institutional reforms take far more than three to four years, and the Bank's project period is too short to actually see reforms through to completion.

Q.2. What aspect of the Bank's assistance—policy advice, lending, analytical work—has contributed the most to the development of agriculture in Africa?

Bank lending was most often indicated (62 percent of responses) as an important contributor to the development of agriculture in Africa, followed by analytical work (50 percent) and policy advice (43 percent). The respondents did not indicate the order of importance.

Respondents offered some interesting views on analytical work:

- Past analytical work has been focused too much on the "standard" situations in which, as always, it has been providing excellent analysis.
- The Bank lacks the courage to draw far-reaching conclusions: a departure from the Green Revolution model as it has worked for the South Asia Region.
- The Bank does not do enough in analytical work. For years none has been done, yet the Bank provides advice freely and develops lending operations based on "borrowed" knowledge.
- Analytical work has helped, but the Bank is in a situation where much of the analytical work done is not used because there are severe limits on funds available for lending.

Q3. What are the Bank priorities for agricultural trade reform in the countries that you know have worked, and has that been clearly communicated to Bank staff working in the Region?

Most respondents noted that there is no clear Bank-wide priority for agricultural trade reform. They believe that the priorities have never been stated explicitly. Also, no clear vision for agricultural trade has been communicated to staff.

At the same time, some respondents believed that:

- The Bank's priorities for agricultural trade reform seem to be mainly to reduce trade barriers and encourage trade in all areas (not just agriculture). There is also emphasis on trade liberalization and elimination of subsidies that is fairly well communicated, but not always accepted by clients.
- Agricultural trade reform has focused on "traditional export commodities" to the exclusion of internal trade in agricultural goods and related inputs, processing, and storage. Unfortunately, since the spate of criticism by international NGOs, the Bank has soft-pedaled support to growth in agricultural exports from low-income countries. At the same time, efforts to liberalize agricultural trade with OECD countries is not likely to get very far.
- The issue is now one of non-tariff barriers, but the Bank is not working on this in a significant way, and there is no teamwork with Poverty Reduction and Economic Management or Development Economics in this area.

Q4. There was a multiple choice question that asked staff to select what should be the top priority for agricultural trade in Africa.

The four options that they were asked to choose from were:

- Promoting measures to increase regional trade
- Promoting reduction in trade barriers and distortions in OECD countries
- Promoting increased production of export crops from African countries
- Promoting measures to achieve food self-sufficiency in African countries.

Thirty-eight percent of the respondents identified promoting measures to increase regional trade among African countries as the top priority for agricultural trade in Africa. This was followed by promoting reduction in trade barriers and distortions in OECD countries (29 percent). Only 13 percent of the respondents identified promoting increased production of export crops from African countries, and 11 percent selected promoting measures to achieve food self-sufficiency in African countries.

Q5. Any other issues not adequately covered in this questionnaire.

The respondents repeated several issues already covered in the questionnaire, but also raised some others.

Issues already covered:

- Adequacy of staff skills in Africa.
- Inadequate analytical work.
- The country dialogue needs to include input from agriculture.
- Agricultural growth is key to reducing poverty.
- Lack of coordination across sectors in the Bank.
- High cost of preparation of agriculture projects.
- Inadequate supervision resources.
- The Country Assistance Strategies are not adequately making the case for agricultural development.
- Inadequate resources for development of agriculture from the Bank and other donors.
- Focus on potential of communities.
- Research and extension.

Additional issues raised by individual respondents:

- Lack of quality control in design and implementation of Bank projects.
- Not enough work is done to verify the feasibility of using a sectorwide approach. Agriculture is multisector; each subsector (for example, credit) is almost a sector. Therefore, a sectorwide approach is unsuitable because it is attempting the impossible.
- There is little understanding in the Bank of traditional farming systems.
- Agriculture is subject to higher standards of evaluation than other sectors.
- The challenge is less to convince people to support agriculture, and more on how to support agriculture.
- Unsuitability of programmatic lending to support agriculture.
- Lack of consistent, sustained project implementation assistance.
- Need to stress the interconnection between agricultural production and industrialization.
- Impacts of droughts and the like in wiping out productivity gains from agricultural growth.

- Organic farming is rising in importance in the industrial countries but is being killed in Africa by the active promotion of chemical-driven farming.
- Transboundary transport infrastructure between countries is key to promoting regional agricultural trade.
- Increasing land titling could improve security of tenure for agribusiness investors.
- Irrigation development should be top priority because of significant rainfall variability and weather risk.
- Difficulty in assessing the impacts of Bank-funded agricultural development activities. The performance indicators that often end up being used (for example, crop yields, value of production, value of agricultural exports) tend to be determined not only by Bank-funded interventions, but also by many other factors outside the control of the Bank.
- The Bank is no longer the dominant force in most of Africa that it once was. Other donors are becoming more important.
- Relationship of the work of the Bank with other global and regional organizations.
- Importance of promoting soil health.
- The issue of food security and its links to rural, human, and general development.

APPENDIX G: SECTOR STAFFING ANALYSIS

Table G.1: Staff Mapped to the Agriculture and Rural Development Department in the Africa Region

Category	1997	1998	1999	2000	2001	2002	2003	2004	2005	2006
Generalists (number)	42	37	32	35	38	37	44	45	43	43
Technical (number)	40	36	35	36	29	28	25	22	20	17
Total of generalists + technical staff	82	73	67	71	67	65	69	67	63	60
Technical staff as percentage of total	48.7	43.9	42.6	43.9	35.3	34.1	30.4	26.8	24.3	20.7
Generalist staff as percentage of total	51.2	50.6	47.7	49.3	56.7	56.9	63.7	67.1	68.2	71.6

Source: Human Resources (HR) Unit of the Bank.

Note: Technical staff included, among other groups, soil scientists and forestry, extension, livestock, agribusiness, and irrigation specialists. Generalist staff included operations officers, economists, and rural development specialists, among other categories.

APPENDIX H: SELECTED EXTENSION APPROACHES

Table H.1: Extension Services

Type of extension service	Origin or characteristics
General national extension services	The standard approach to public sector extension with field advisory services provided free to farmers throughout the country.
General agricultural extension	The traditional form of extension that has been dominant for the past 80 years.
Training and visit extension (T&V)	Debuted in the late 1960s as a reform of ineffective general extension services.
Strategic Extension Campaign (SEC)	Methodology developed by FAO to systematically incorporate peoples' participation into a national extension program.
Extension by educational institutions	Especially for agricultural universities, can be the dominant approach to national extension.
Publicly contracted extension	Services are provided by private firms or NGOs on contract to government.
Targeted extension services	Some extension approaches attempt to avoid the high recurrent costs by narrowing their focus in subject matter, clients, region, or time.
Specialized extension services	Focus efforts on improving production of a specific commodity or aspect of farming (such as irrigation, fertilizer use, forest management, and the like).
Project-based extension	Focus increased extension resources on a defined area for a specific period of time.
Client-group-targeted extension	Focuses on specific types of farmers, usually on disadvantaged groups, such as small farmers, women, minorities, or disadvantaged ethnic groups.
Producer-led extension services	These approaches involve farmers in the work of extension—drawing on producers' knowledge and resources.
Animation Rurale (AR)	Introduced in francophone Africa as a strategy to break the top-down pattern found in most development programs.
Participatory extension	Harnesses farmers' own capacities to organize group meetings, identify needs and priorities, plan extension activities, and use indigenous knowledge to improve production systems.
Farming systems development extension	Requires a partnership between extension, researchers, and local farmers or farmer organizations.
Producer-organized extension services	Completely planned and administered by producers.

(*Continues on the following page.*)

Table H.1: Extension Services (continued)

Type of extension service	Origin or characteristics
Commercialized extension services	These approaches rely on commercialized extension.
Cost-sharing extension	May be incorporated into any of the other extension approaches by requiring farmers to share costs of services.
Commercial extension advisory services	Are becoming more common, as the rationale for free public extension services is questioned and farmers find they need more dependable or specialized services than are available from a public extension agency.
Agribusiness extension	Supports commercial interests of input suppliers and produce buyers who require or benefit from provision of sound extension services to support farm production and management.
Mass media extension	These approaches support other extension efforts or provide information services to a general audience.
Mass media extension	Provides pure information services directed to a wide audience.
Facilitated mass media	Links mass media information services with field extension agents or farmer-extensions to facilitate discussion and understanding of issues.
Communications technologies	Allow people in rural areas to interact with specialists or specialized sources of information through rural telephone or internet services possibly institutionalized in "telecottages" for community access.

Source: World Bank 2002b.

APPENDIX I: COTTON SECTOR REFORMS: AN UNFINISHED STORY

Cotton is critical to the economic development of several countries in West Africa (Benin, Burkina Faso, Chad, Mali) and East Africa (Mozambique, Tanzania, Zambia, Zimbabwe). It is often considered a success story because between 1980 and 2000, while Africa's share of world agricultural trade fell by half, its share of world cotton trade rose by 30 percent, and cotton production was able to contribute significantly to poverty reduction in some countries, such as Burkina Faso. This was mainly because cotton is predominantly a smallholder crop. Over 2 million poor rural households in Africa depend on it as their main source of cash income (Tschirley and others 2006a). Cotton cultivation has also made possible growth in infrastructure and greater satisfaction of basic needs such as health and education in some countries. However, dependence on a single export crop has also made smallholders in many countries vulnerable to world prices.

Before the adjustment era, the marketing and trade of cotton in most African countries was handled by parastatals, which in several cases also met the input and credit needs of the farmers. The Bank has provided considerable support for cotton sector reform in the Region for the past 10 years. Though the specific reforms undertaken have varied according to country circumstances, the broad goals of the reforms have been similar: to improve the efficiency and competitiveness of the sector.

In several countries the Bank has provided support for privatization of the parastatals, linking producer prices to world markets, ginnery rehabilitation, improving grading practices, research in and adoption of new varieties of cotton, and strengthening the capacity of producer organizations to play an increasing role in management of the cotton sector, among other reforms.

The cotton reform story is unusually complex, because neither the Bank nor its clients in Africa are in a position to influence cotton-production subsidies in the United States and other developed countries. The subsidies in the developed countries have increased production and consequently depressed world market prices. Whether the removal of subsidies would actually lead to higher world prices for cotton is debatable (since U.S. exports would likely be replaced by those of higher-cost producers), but research points to considerable revenue forgone by African countries because of these subsidies (World Bank 2006e). In addition, pest management techniques and technology improvements that contributed to increased yields have reduced production costs in major world producers such as the Brazil, China, and the United States, making it difficult for African countries to compete.

It is difficult to draw conclusions about the outcome of Bank interventions, partly because it is difficult to trace causality. Also, reforms have been implemented at differing paces and to different degrees across countries. For example, in Zambia the government completely liberalized the cotton sector, whereas in Mali the privatization of the main parastatal has not yet taken place. Though there have been organizational differences in structure and pricing policies in the cotton industry among the various countries, there have been common technical challenges in maintaining quantity and quality of production in the face of declining and highly volatile world prices. While in some countries, such as Burkina Faso, organizations of producers have taken on major responsibility for a growing number of functions in the sector, this has not happened across countries. Some gains that have followed the reform period include a higher percentage of market prices for farmers, more timely payments, and reduced pressures on state budgets. However, with the privatization of the parastatals, the private sector has not stepped in to fill the gap left in the supply of inputs and credits.

The cotton sector faces the same constraints as other crops do because of the reform process: lack of access to inputs (fertilizers, pesticides, seeds), extension, and credit. The Bank's approach to cotton sector reform in Africa does not show adequate recognition of how the sector had been insulated from some of these problems because of the special role played by parastatals in input supply and credit access. Data show that cotton yields have stagnated in most countries—including Benin, Burkina Faso, Chad, and Tanzania. Lack of inputs and declining soil fertility (particularly because in several countries expanded output under cotton production resulted from increasingly marginal areas being brought under cultivation for the crop) remain major concerns. Tschirley and others (2006a) also note nine technical challenges the sector faces (box I.1). The Bank is now at a crossroads. Given its long-term involvement in the sector, other donors and clients are looking to the Bank for advice in how to move forward.

Despite its long involvement in the sector, the Bank has not—until very recently (and ongoing)—attempted to undertake rigorous analytical work that identifies the multiple constraints to development in the cotton sector and lessons of experience across the Region to inform its policy dialogue with the clients.

Box I.1: Technical Challenges in the Cotton Sector

- **Support strong varietal research and dissemination.** Seed quality has major impacts on yields, ginning ratios, and fiber characteristics. It thus establishes the outer limits of productivity and quality throughout the system.
- **Maintain the purity of varieties once they are released.** This typically requires varietal zoning agreements, which demand some level of horizontal coordination among players.
- **Assure sufficient and timely provision of treated seed to farmers.** Treated seed reduces disease in a very cost-effective manner.
- **Ensure sufficient and timely provision of appropriate pesticides to farmers.** Most cotton varieties currently in use in Africa are highly susceptible to attack by pests, so that in many areas three to five pesticide applications are considered necessary for economical yields.
- **Manage pesticide use to reduce cost and avoid insect resistance.** The "pesticide treadmill"—inappropriate use of chemicals that increases insect resistance, leading to more use—increases financial costs and both environmental and human health externalities.
- **Manage pesticide use to reduce damage to human health and the environment.** This issue has received very little attention to date, and is becoming increasingly important within several francophone systems. Maumbe and Swinton (2003) note the significant health costs incurred by pesticide-using cotton farmers in Zimbabwe.
- **Ensure appropriate use of fertilizers.** High cost of fertilizers and varieties that do not respond well to fertilizer means that this input is often not profitable for cotton in Africa. Wider use, which may be a prerequisite for cotton to make major and sustainable contributions to poverty reduction, requires reducing its cost and combining it with improved varieties that are more responsive to fertilizer.
- **Control quality from the farm gate through the export of fiber.** Quality relates to fiber characteristics and to the uniformity of these characteristics in any given export lot. Countries with a reputation for high and uniform quality will have a ready market and better prices for their output, even during the periodic gluts that afflict the world cotton market.
- **Pay farmers sufficiently remunerative prices to ensure their continued and increasing participation in the sector.**

Source: Tschirley and others 2006a.

APPENDIX J: MARKETING REFORM

Table J.1: Some Examples of Policy and Market Reform from the Portfolio Review

Approval fiscal year	Country	Project name	Project ID	Lending instrument type	Planned reform
1991	Zambia	Recovery Credit	P003235	Adjustment	Decontrol of maize and fertilizer marketing and pricing; privatization of all parastatals except public utilities and natural monopolies; trade liberalization, involving tariff reform, the removal of export restraints.
1992	Tanzania	Agricultural Adjustment	P002818 (P002776 Parent Project)	Adjustment	Reform the pricing and marketing systems of food crops and three major export crops—coffee, cotton, and cashew nuts. Restructuring of crop processing facilities.
1992	São Tomé and Principe	Agriculture Sector	P002535	Investment	Privatize the publicly owned agricultural estate: • Distribute and lease a major part of the Public Agricultural Enterprises (Empresas Estatal Agricola, EEAs) to smallholders and medium-size farm and agro-processing enterprises. • Reduce the number of estate laborers and increase labor productivity on the remaining (private) estates. • Lease the financially viable EEA and nucleus-processing facilities to the private sector. • Reduce export taxes on cocoa.
1992	Mozambique	Economic Recovery Credit (ERC)	P001775	Adjustment	Foreign exchange system reform. Agriculture price reform: • Adjust the floor prices of cotton and cashew in line with the evolution of border prices. • Remove policy constraints preventing traders from operating in rural areas. • Review the role of AGRICOM (the state marketing agency). • Privatize Caju de Mozambique (the largest state-owned processing enterprise in the cashew sector).
1993	Zimbabwe	SAC II	P003322	Adjustment	Agriculture market reform.
1993	Malawi	Agriculture Services	P001660	Investment	Increase the availability of improved seeds and fertilizers to smallholders by supporting the formulation and implementation of seed and fertilizer policy reforms and financing incremental fertilizer.

Result
All price controls have been abolished. Producer prices for all crops are set by supply and demand (although there is still a producer floor price for maize). All prices were decontrolled in 1989, except maize meal and fertilizer prices. Private traders can buy and sell all agricultural products with no public monopolies. The NAMBOARD (National Ag Marketing Board) structures have been disbanded. Fertilizer importation and marketing are fully liberalized.
Grain marketing and pricing policy. The expected private sector investment in grain marketing business, including construction of warehouses, did not take place. All food crops are now freely marketable. While the government no longer determines producer prices, it does establish the SGR floor price for the purchase of grain for food security reserves. In contrast to the grains subsector, where reforms were well under way, the reforms related to the export crops subsector had only recently commenced. The project was successful in starting the withdrawal of parastatals from agricultural production; introducing competition in the supply of seeds and fertilizer; rationalizing and substantially reducing the number of agricultural projects; and providing continuing support for agricultural policy analysis and project management. The Tanzanian Seed Corporation, TANSEED, was reorganized, and seed companies were established by early 1991. With respect to fertilizer, reforms were initiated slowly, because the government failed to raise fertilizer prices to the agreed level by September 1990. But by June 1991 the prices had been raised to the agreed targets. In addition, the subsidy was made explicit.
Two private NMCs, SODEAP and SAC Sur, were created in order to increase competition with the already established private enterprises managing or leasing the rehabilitated estates for the purchase of cocoa and the provision of inputs and credit. The NMCs are still operating in the northern and southern part of the country but are only partially fulfilling their mandate. They have been purchasing and processing smaller and smaller amounts of cocoa over the past two years and stopped providing seasonal credit some years ago because of reduced access to working capital and farmers' very poor repayment rates. In addition, four of six private enterprises operating at the beginning of the project cancelled their leases because of labor problems and poor results (yield forecasts for the cocoa replantations in Uba Budo and Sta. Marguerida were 1,500 kilograms of dried cocoa per hectare, but in reality only about 700 kilograms were obtained on the best plots, with a general average of only 350 kilograms). The reason for the poor results was the introduction of inadequate planting hybrid materials during CRP. Following their departure, the government asked the project to distribute the land of these estates. But as a consequence, the quality of the marketing, input supply, and credit services provided to farmers has been declining dramatically.
Floating exchange rate policy was adopted. The official exchange rate was set on the basis of the parallel market exchange rate. The objectives of the ERC were met, but after some delays. Private sector participation in the domestic marketing of agricultural products increased sharply with the relaxation of the licensing requirements on retailers and wholesalers. Prices of agricultural commodities rose above the minimum prices, which rendered obsolete the envisaged review of minimum prices and AGRICOM, the state marketing board, whose share in the procurement of maize declined drastically. The main domestic effects of the policy measures were to increase agricultural marketing, particularly for maize, and to reverse the worsening of the terms of trade between agriculture and industry in regions where small private traders were active. Externally, the policy measures led to increased exports.
Liberalization of trade and exchange rate by progressively moving to a unified, market-based foreign exchange system and an import regime based on modest, tariff-based protection. Removal of price controls in beef, dairy, cotton, yellow and white maize, oilseeds, and wheat and elimination of marketing board monopolies. Slaughter quotas imposed by the Department of Veterinary Services were eliminated so that the private sector could participate more actively in meat processing. The turnaround of the Grain Marketing Board was one of the most important public sector financial management improvements under SAC II. Deregulation was not complete, however.
The reform with the most far-reaching implications was the amendment to the Special Crops Act, which allowed smallholders to begin growing burley tobacco. This, together with support targeted at burley groups by SFSP, resulted in a major expansion in the number of smallholders growing burley from 18,000 to 50,000, and substantially enhanced incomes for these farmers. The deregulation of fertilizer imports has been partly achieved, and at one stage nine local and international companies were active. Subsidies, which started to be reduced from 1991/92, were completely removed in 1993/94. Toward the end of the project, however, only two private fertilizer companies were operating, and the government has again been playing a major role in fertilizer importation, reversing the liberalization trend of the fertilizer policy. There has been no success in reducing the cost of fertilizer imports to farmers for a number of technical and policy reasons.

(Continues on the following page.)

Table J.1: Some Examples of Policy and Market Reform from the Portfolio Review (continued)

Approval fiscal year	Country	Project name	Project ID	Lending instrument type	Planned reform
1993	Malawi (continued)				
1993	Kenya	Parastatal Reform Technical Assistance	P001348	Investment	Restructuring, preparation for privatization, and commercialization of specific parastatals (Kenya Tea Development Authority, National Cereals and Produce Board [NCPB]).
1994	Chad	Economic Recovery	P035594	Adjustment	Increase producer price of cotton by 50 percent.
1995	Ethiopia	National Fertilizer Sector Project	P000753	Investment	Decontrol retail and wholesale fertilizer prices. Eliminate fertilizer subsidies. Develop institutional mechanisms to ensure that both public and private sector importers would have equal access to IDA and government funds for importing fertilizers. Level playing field between fertilizer distributors by eliminating special access to government-owned warehouses by the state-owned Agricultural Inputs Supply Corporation (AISCO).
1995	Benin	SAC III	P000111	Adjustment	Divesture of public agro-processing companies. Cotton sector reforms: • Transfer of SONAPRA, the cotton company, into a mixed capital company. • Adopt market-based pricing mechanism procedures for seed cotton sales to private gins and revise the price stabilization mechanism.
1998	Cameroon	Cameroon - SAC III	P054443	Adjustment	Privatization of agro-industries (palm oil, cotton, sugar, and fruits).
1998	Lesotho	Agriculture Policy and Capacity Building	P001402	Investment	Introducing changes in management through institutional restructuring, privatization, and divestiture of activities and market liberalization.

Result
The objective of liberalizing production and marketing of hybrid seed was achieved, and all subsidies on improved seed were removed. Two commercial companies are currently producing or importing almost all hybrid maize seed used. Overall, while the agreed policy reforms have been largely implemented, this component has not fully achieved its objectives because the overall impact on the availability of inputs to small farmers, and competitiveness in supply, has been very modest. The policy reforms, in particular the liberalization of the markets for hybrid seed and burley tobacco, have encouraged greater involvement of private seed and fertilizer companies, but at the time of project closure there were clear signs that the government is becoming increasingly involved in fertilizer distribution again.
Twenty-nine tea factories were sold to tea farmers. National Cereals and Produce Board (NCPB). A contract with Agriconsult of Australia was reached as part of an agreement under an agricultural adjustment credit that the government would commercialize the entity and get out of the business of managing a strategic reserve. The government's contract with Agriconsult allowed the advisors to sell off the silos and to undertake a retrenchment program that has improved the environment for grain production in the country. The project had a positive impact on some small enterprises that were privatized and became more efficient in their operations. However, many of the small firms—for example, the ginneries—were sold by what were considered nontransparent processes to parties that have ceased to operate them because of insufficient investment funds. The result has been that the cotton ginneries have become a major bottleneck and a major reason for the dramatic decline of the cotton sector in Kenya.
Increase in producer prices of cotton by 50 percent for the 1994-95 crop season and reinstate the previous cotton-sharing system. The liberalization of most prices and the increase in the producer price for first-grade cotton have enhanced producer incentives (as reflected in increased cotton plantings), while reduced inflation has fostered the emergence of a sounder economic environment; more effective customs operations have been established, and the program has brought down the government's accumulation of payments arrears and absorption of available credit, allowing it to improve the liquidity of the private sector.
All agreed policy reforms were fully carried out, but the objective of the reforms—creation of a functioning and competitive industry—was not accomplished. Fertilizer pricing was totally liberalized, and fertilizer subsidies abolished. A fertilizer trade and manufacturing proclamation was issued in 1998, which set fertilizer standards and enabled the government to start enforcement of fertilizer quality standards from port to retail. Further, in order not to have an unfair advantage over its competitors, the government parastatal AISCO withdrew from marketing centers supported by Ministry of Agriculture staff, and ceased to have preferential access to the ministry's warehouses. Finally, foreign exchange for fertilizer importation was allocated among importers in a fair and transparent manner throughout the life of the project. However, government-introduced programs (already discussed under Component 2 above), although well-meaning in their intentions, had a design that was deleterious to competitive market development. Furthermore, more could have been done to address the persistent allegations of privileged market access by some regional trading houses.
Cotton production more than doubled in the 1990s. But quality of inputs distributed to farmers by some private suppliers was less than adequate. As a result, production was expected to decline in the 1998/99 crop year. The farm gate price was increased from 80 CFAF/kg prior to the devaluation to 200 CFAF/kg in the two crop seasons 1996–98, but the producers' share in SONAPRA's after-tax profits has not been adjusted, and represents only a small portion of their income (less than 2 percent). Their share in cotton exports (a proxy for the industry's revenues) actually declined during the decade, from 63 percent in 1991–93 to 55 percent in 1996–98. The allocation of seed cotton among private ginning companies remains an administrative decision.
Privatization of agro-industries partially met the targeted objectives. The outcomes envisaged under this subcomponent were met for sugar, palm oil, and tea, but not for the other crops.
Agreed policy statement on subsidies for farm inputs based on subsidy evaluation review. Cabinet has issued a policy directive on the use of input subsidies. Deregulation of remaining controls on agricultural commodities implemented according to schedule. Deregulation schedule has been submitted to Cabinet for approval. Privatization, deregulation, and liquidation have not progressed as envisaged at appraisal. Of the 16 enterprises identified to be privatized at appraisal, 1 has been privatized (leased); 4 liquidated; with a further 2 being partially liquidated. Project activities related to marketing facilitation/reform have not yet led to deregulation. However, the project succeeded in carrying out studies, including an analysis of the 13 commodities that are under government control, with the view to understanding the impact of liberalization/deregulation on producers, consumers, and trade in general. The results of these studies were discussed with stakeholders in workshops carried out in all 10 districts, the recommendations of which were discussed in a national workshop in March 2003.

(Continues on the following page.)

Table J.1: Some Examples of Policy and Market Reform from the Portfolio Review (continued)

Approval fiscal year	Country	Project name	Project ID	Lending instrument type	Planned reform
1999	Rwanda	Economic Recovery Credit	P057294	Adjustment	Increase in tea prices and the removal of the coffee tax. Adoption of a comprehensive strategy to revive the coffee sector; privatize tea factories and estates and establish stakeholder-based regulatory frameworks for the tea and the coffee subsectors. Privatization of tea factories and participation of tea farmers in ownership of factory. Adoption of market-oriented policy framework for distribution and marketing of agricultural inputs.
2005	Tanzania	Tz-PRSC2	P074073	Adjustment	Review the role of crop boards to limit their functions to regulatory activities.

Note: Because of the problems with reporting and attribution (as discussed in the section on M&E), it is not always possible to determine the outcome of Bank interventions. The above list includes cases where it was possible to determine achievements based on the information provided in project completion reports.

Result
In 1999, the government eliminated the 30 percent tax on coffee exports and increased the producer price of tea by 37 percent. Another important development has been the emergence of producers' associations that have become active in selling coffee directly to exporters and in distributing inputs such as fertilizers and pesticides to members. Legislation was passed in 2000 to change the legal mandates of OCIR-Cafe and OCIR-The, the two parastatals involved in production, marketing, and regulatory functions in the coffee and tea sectors, limiting their role to regulation, monitoring, and promotion. The privatization of the tea factories has not yet taken place. Most of the coffee-processing plants have been privatized and the privatization of the nine state-owned tea estates is expected to take place. In line with its policy of liberalization of markets, the government has reaffirmed the policy of market-based pricing and distribution of these inputs, thus abandoning the pre-genocide practice of state control of the market for these inputs.
The review work is being done in phases. Initial work on the review process began in September 2003 with an institutional mapping exercise of coffee, cotton, cashew, and tea, followed by funding, institutional, and impact evaluation of the four Crop Boards.

Table K.1: Current and Potential Irrigated Area in Africa and Selected Countries

Country	Cultivated area ('000 ha)	Irrigated area ('000 ha)	Potentially irrigable land ('000 ha)	Current irrigated area as percent of cultivated area (%)	Current irrigated area as percent of potential (%)
Ethiopia	10.671	290	2.700	3	11.0
Somalia	1.071	200	240	18.7	83.3
Madagascar	3.550	1.086	1.517	30.6	71.6
Sudan	16.653	1.863	2.784	11.2	66.9
Zimbabwe	3.350	174	366	5.2	47.5
Mali	4.700	236	566	5.0	41.7
Malawi	2.440	56	162	2.3	34.8
Zambia	5.289	156	523	2.9	29.8
Nigeria	33.000	293	2.331	0.9	12.6
Mozambique	4.435	118	3.072	2.7	3.8
Ghana	6.331	31	1.900	0.5	1.6
Kenya	5.162	103	353	2	29.0
Tanzania	5.100	184	2132	3.6	9.0
Congo, Democratic Republic of	7.800	11	7.000	0.1	0.2
Total, Sub-Saharan Africa	182.682	7.105	39.413	3.9	18.0
Total, Sub-Saharan Africa excluding the three largest irrigation countries	146.767	2.658	33.613	1.8	7.9

Source: Peacock, Ward, and Gambarelli 2007.

From Conference Paper No. 8
"New Challenges in the Cassava Transformation in Nigeria and Ghana"
By Felix Nweke

Nature of the Success

Why is it considered a success?

- Production triples within a decade, from 1984 to 1992
- Nigeria surpasses Brazil as world's leading cassava producer
- Sixty percent of Nigerian villages plant improved varieties
- Resulting price fall benefits consumers, making cassava a powerful poverty fighter.

Motors of change

- Improved varieties (tropical Manioc Selection (TMS)): high yielding, early bulking, and disease resistant
- Biological control of mealybug epidemic
- Processing technology development: gari (dried prepared cassava porridge), mechanical grater to release processing labor
- Change from inhibiting to favorable trade policies.

What constrains further expansion?

- Harvesting labor bottlenecks
- Market competition from subsidized imported starches.

Aggregate Impact

Scale and productivity gains

- Five million farmers produce cassava
- Cassava accounts for 12 percent of farmers' cash income.

Equity

- Broad access to improved varieties across farm sizes
- Cash production concentrated, 50 percent among top 10 percent of households, but less concentrated than maize (70 percent cash sales among top 10 percent of farm households)
- Poor consumers are major beneficiaries of a 30-year productivity-induced fall in real cassava and gari prices.

Sustainability

- Financial: highly profitable for smallholders, returns to HYV plus mechanical grating 20 times greater than traditional varieties with hand grating
- Ecological: long-term yields sustainable without fertilizer.

Lessons for Building Future Successes

Resume long-term funding for cassava research

Processing technology necessary for rapid market development

Table L.1: Dynamics and Drivers for Change

Timing	Phase 1 Cassava becomes a staple food, 1910–60	Phase 2 Laying the foundation, 1960–77	Phase 3 Mealybug invasion, 1978–83	Phase 4 The surge, 1984–92	Phase 5 New challenges, 1993 on
Key actors	Immigrants Farmers	Rural artisans IITA Shell Oil	IITA	Government National Root Crop Research Institute Private oil companies	
Motors of change	Severe rural labor shortages (the result of wars and influenza epidemic of 1918) induce a move out of labor-demanding cocoyam and into cassava. Emancipated slaves from Sierra Leone introduce gari processing technology. Immigrants bring in new, bitter varieties.	Mechanical graters imported from Benin and refined by local artisans. Graters spread, releasing processing bottlenecks. TMS varieties developed (1971–77) but fail to spread rapidly.	Mealybug invasion attacks cassava crop.	Biological control of mealybug (1981 on) takes effect. Policy changes stifle food imports —drop food import subsidies —ban on cereal imports —devaluation of the naira raises food import prices. Government includes cassava in extension programs. Oil companies help finance cassava promotion.	Rising wage rates lead to labor constraints in harvesting and processing. Imported corn starch becomes cheaper than cassava starch.
Beneficiaries	Small farmers Urban gari consumers	Small farmers Urban consumers		Cassava farmers Urban consumers	
Production gains	Production doubles from 1948 to 1958	Grater induces 50% increase in production. Annual growth 2.5% per year.	Production falls 20% –3.7% per year.	Production increases 150%. Annual growth rate of 12% per year.	Production up 15% Annual growth rate slows to 1.5% per year.
Impact	Cassava becomes established as a rural food staple. Growing urban markets attract gari trade.		Massive mobilization for biological control of mealybug across Africa.	Real gari prices fall. Gari/yam price ratio falls by 50%. Gari/rice price ratio falls by 25%.	Consumer gari prices trend upward. Industrial demand for cassava starch stalls.

APPENDIX M: PREPARATION COSTS AND RISK RATINGS FOR AGRICULTURE PROJECTS

Table M.1: Africa Region Projects: Average Preparation Costs over Time (non-agriculture versus agriculture)

Approval fiscal year	Non-agriculture projects (US$)	Agriculture projects (US$)
1991	325	711
1992	219	2,049
1993	242	1,069
1994	260	1,148
1995	288	1,551
1996	370	3,459
1997	313	929
1998	308	6,734
1999	254	1,841
2000	234	
2001	252	3,149
2002	230	5,762
2003	261	2,872
2004	289	2,861
2005	391	2,628
2006	360	3,145
Overall results	288	1,969

Source: World Bank data.

Table M.2: Africa Region: Projects at Risk over Time (percent, all projects versus agriculture projects)

	At risk (%)	
Fiscal year	All projects	Agriculture projects
1991	54.9	65.3
1992	49.4	57.8
1993	47.2	55.4
1994	48.1	52.3
1995	44.4	48.1
1996	34.8	24.5
1997	39.5	33.0
1998	30.8	24.4
1999	27.3	17.5
2000	13.9	18.4
2001	14.8	14.5
2002	26.2	32.9
2003	19.0	17.6
2004	22.8	22.0
2005	29.0	39.1
2006	21.9	23.0

Source: World Bank data.

Management Response

1. See *World Development Report 2008: Agriculture for Development, Discussion Draft,* May 21, 2007. The final version of the WDR is scheduled to be launched on October 19, 2007.

2. IEG notes that its review presents country-specific data to show the differences in performance among three categories of countries: the comparatively better performers, the medium performers, and the poor performers. Aggregation of growth rates for agriculture for Sub-Saharan Africa as a whole presents challenges given the wide variation in the rates and sources of growth across countries (see chapter 2 and appendix table B.2 of the IEG report).

3. IEG notes that some recommendations of its study that are critical for the development of agriculture in Sub-Saharan Africa, such as the importance of increasing the productivity of rain-fed agriculture, have not been addressed in the Management Action Record.

Management notes that it fully agrees with the importance of increasing the productivity of rain-fed agriculture. The elements of the actions it is undertaking in supporting comprehensive agricultural programs at the country and regional levels as noted in the Management Action Record are defined to accelerate growth and productivity. As is now the case, most improvements will be achieved in rain-fed areas, even with the planned expansion of irrigated areas.

Réponse de la direction

1. Voir le projet de document « Rapport sur le développement dans le monde 2008 : l'agriculture au service du développement », du 21 mai 2007. La version finale du RDM est prévue pour être publiée le 19 octobre 2007.

2. L'IEG fait remarquer que son rapport présente des données par pays pour montrer les écarts de performances entre trois catégories de pays, classés selon leurs performances (comparativement meilleures, moyennes et médiocres). Il est difficile de faire la somme des taux de croissance agricole pour l'ensemble de l'Afrique subsaharienne dans la mesure où ces taux et les origines de la croissance varient de façon importante d'un pays à l'autre (voir le chapitre 2 et le tableau B.2 en annexe au rapport de l'IEG).

3. L'IEG fait remarquer que certaines recommandations de son étude qui ont une immense portée sur le développement de l'agriculture en Afrique subsaharienne, par exemple l'importance d'augmenter la productivité de l'agriculture pluviale, ne sont pas mentionnées dans le Relevé des interventions de la direction.

La direction se déclare entièrement d'accord sur l'importance d'augmenter la productivité de l'agriculture pluviale. Les éléments des actions qu'elle est en train d'engager pour appuyer des programmes agricoles intégrés au niveau des pays et de la région, tels qu'indiqués dans le Relevé des interventions de la direction, sont définis pour accélérer la croissance et la productivité. Comme c'est le cas aujourd'hui, la plupart des améliorations seront obtenues dans des zones non irriguées, même avec l'extension prévue des périmètres irrigués.

A Resposta da Administração

1. Vide Relatório Sobre o Desenvolvimento Mundial de 2008: A Agricultura para o Desenvolvimento, Versão Preliminar para as Deliberações, 21 de Maio de 2007. A versão final do RDM deverá ser divulgada em 19 de Outubro de 2007.

2. O GIA assinala que a sua análise apresenta dados específicos aos países para mostrar a diferença do desempenho em três categorias de países: aqueles que tiveram um desempenho comparativamente melhor, aqueles que tiveram um desempenho médio e aqueles cujo desempenho foi fraco. As taxas agregadas de crescimento da agricultura na África Subsariana, como um todo, apresentam desafios que são devidos à grande variação das taxas e fontes de crescimento através dos pa-

íses (vide o capítulo 2 e o apêndice no quadro B.2 do relatório do GIA).

3. O GIA assinala que algumas das recomendações do seu estudo que são cruciais para o desenvolvimento da agricultura na África Subsariana, tais como a importância de aumentar a produtividade da agricultura de sequeiro, não foram abordadas no Registo das Acções Empreendidas pela Administração.

A Administração diz que concorda plenamente com a importância de aumentar a produtividade da agricultura de sequeiro. Os elementos das acções que está a empreender para apoiar os programas abrantentes da agricultura a nível dos países e a nível regional, como indicado no Registo das Acções Empreendidas pela Administração, são definidos em termos de acelerar o crescimento e a produtividade. Como é o caso actualmente, a maioria dos melhoramentos serão alcançados nas zonas de sequeiro, mesmo com a ampliação planeada das terras irrigadas.

Chapter 1

1. Using a poverty line of US$2.15 per day.

2. "But the decision of Africa's new leaders to invest in industry in isolation from village agriculture and rural industries was also consonant with the views of many Western development economists in the 1950s, who assumed that agriculture was a passive sector, a black box that could be squeezed to finance industry" (Eicher 1999, p. 17).

3. "In Africa, instead of focusing on food production and the building of the basic institutions for a modern agriculture over a period of decades, donors jettisoned much of what had been learned about the agriculture development experience in Asia and in the seventies introduced new programs, such as a diffused provision of services targeted to the poor, integrated rural development, programs targeted to women and an attack on environmental problems. These programs overlooked the critical need to address concurrently agricultural productivity and sustainability issues" (Mellor 1998, p. 59).

4. "Because most land is held communally in Africa, in most cases rural inequality does not stem from severe inequality in landholdings. Rather it reflects geographic differences in the quality of land, in climatic conditions, and in access to markets and to remittances from urban areas" (World Bank 2000, p. 93).

5. Many rural and urban poor in Africa are net food buyers (Christiaensen and Demery 2007).

6. Diao and others (2006) report similar findings.

Chapter 2

1. Similarly, in Mozambique, the growth rate jumped in the 2000–04 period because of the post-conflict resettlement of refugees in the rural areas and the resulting expansion in production (World Bank 2006g).

2. Management agrees that agricultural growth and productivity must increase relative to current levels, and notes that the growth rate of agricultural GDP in Sub-Saharan Africa in the period 1980–90 averaged 2.3 percent annually. This rate increased to 3.3 percent annually between 1990 and 2000, and to 3.8 percent annually between 2000 and 2005 (World Development Indicators, 2007). IEG notes that these numbers mask substantial variation across countries and over time.

3. A background paper produced for the Commission for Africa Report (2005) found that international efforts for harmonizing disbursement and procurement procedures among donors and aligning them more closely with the procedures of African governments have improved coordination. While these changes are welcomed by African governments, they are also seen as focusing more on harmonization of procedures than on aligning donor policies with those of the strategies of African governments (Johnson, Martin, and Bargawi 2004).

4. "We will no longer look only at irrigation and drainage, but also at water resource allocation and comprehensive management. We will not deal with agriculture, forestry, or livestock separately, but with the management of natural resources in sustainable production systems. We will look at rural entrepreneurship instead of agricultural credit, off-farm employment, agro-industries and marketing in isolation. And we will integrate human capital development, infrastructure, and social development into rural development strategies and programs" (World Bank 1997c, p. 17).

5. Agro-ecological zones share similar soil, landform, and climatic characteristics.

6. Management notes that Africa's physical endowment is more favorable for agriculture than this characterization suggests.

7. "That diversification is partly a response to climatic risk is shown by differences in the extent of crop diversification between ecological zones: in the humid forest areas where rainfall is reliable households often are highly specialized, growing only one or two crops. Similarly, households living in the Sahelian zone in West Africa are more diversified than households in areas with more reliable rainfall" (Collier and Gunning 1997, p. 15).

8. Millet is grown in difficult agro-ecological situations (low rainfall, high temperatures, and degraded soils) where maize and sorghum production may not be possible or as productive. Millet is also able to access water from much lower in the subsoil than maize and sorghum. This means that if nitrates are leached beyond the effective depth of a sorghum root system (a common occurrence in the semi-arid tropics), millet plants may still be able to use these nitrates (Yanggen and others 1998).

9. Also, fertilizer response declines as soil health, especially organic content, declines.

10. Four countries in Sub-Saharan Africa have average intensity of fertilizer use greater than 25 kilograms per hectare: Kenya, Swaziland, Malawi, and Zimbabwe. Kenya in particular has experienced tremendous growth in fertilizer use since the early 1990s (Ariga, Jayne, and Nyoro 2006).

11. Research has opened a range of intensification options for individual farmers. These lie along a continuum from adoption of extensive farming only (if surplus land is available) to a low-input sustainable approach (that uses minimum tillage, labor-intensive recycling of nutrients by alley cropping, green manuring and composting, and little or no fertilizer, pesticides or herbicides) to a high-input farm capital intensification (including fertilizers, pesticides, and the like) approach. An individual farmer could be anywhere on the continuum, depending on his individual circumstances, including: access to land and extension, education and tenure arrangements, the need to spread risk, and access to inputs and credit, among other factors.

Chapter 3

1. A review of Bank data indicates that in some countries only one piece of AAA is conducted every year, and in some years none is undertaken.

2. "In the cases of PERs, for example, which are supposed to pull together themes for various sectors, the technical input from ARD sector specialists is often limited to reviewing drafts at a late stage when such inputs are least effective. Even in countries in which the Bank is heavily involved in areas such as rural infrastructure, rural education and health, integration of ARD sector work was often hard to detect" (QAG 2004, p. 6).

3. QAG reports also note weak linkages with lending operations, "sector studies are frequently undertaken to justify/support operations in advanced stage of preparation instead of preceding such preparation efforts." (QAG 2004). The same QAG report also points to "limited impact of the analytical work on the client and the Bank due to shortcomings in dissemination, the short 'shelf-life' of reports, and inadequate systems for archiving, updating and accessing AAA reports" (p. 3).

4. QAG review of Quality of Country AAA (QAG 2005) notes that "Both an analysis of the country assessments and an in-depth review of eight countries that were identified because of their importance for the Bank's program in the Agriculture and Rural Development (ARD) area, suggest that there is an under investment in analytic work in this sector relative to its importance for poverty reduction. The gap was most pronounced in the cases of the Africa and South Asia Regions" (p. 71).

5. "The Bank is slow to revise country AAA in response to political, economic or other changes within the country" (QAG 2005, p. 35).

6. A QAG assessment of the ARD AAA program noted "the Bank appears to have a very porous institutional memory and an entirely inadequate filing system (electronic or otherwise). Many reports were missing or could only be located with difficulty." The study urged ARD to make "greater effort to accurately record AAA tasks in SAP [a Bank data management system] and to ensure that reports are properly archived and readily available to the staff and the client." (QAG 2004, p. 12). An ARD annual retrospective of agriculture and rural analytical work also found that of 186 agriculture and rural ESW reports completed between fiscal 2000 and 2004, only slightly more than half were available in the Bank's internal database and ImageBank.

7. Internal reports have also noted that the Bank is missing opportunities to disseminate analytical work of potential interest to clients, particularly by not translating reports into local languages and by not formally publishing them.

8. "Sector 'silos' are very apparent in the AAA work program, with little or no evidence of interaction between sector departments" (QAG 2004, p. iii; QAG 2005, p. 35, has a similar finding).

9. The dollar amount assigned to agriculture in a development policy loan is not a meaningful number as it is based on sector assignments by task teams.

Chapter 4

1. "There was a significant reduction in the numbers of technical staff, including particularly valuable individuals with long experience for whom no position could be found at the grade to which they had previ-

ously been promoted on the basis of their performance. These senior professionals were offered a choice of either a position at a lower grade, or substantial monetary compensation if they chose to leave the Bank. This exacerbated the attrition of very experienced technical staff resulting from normal retirement" (World Bank 1991a, pp. 4–5).

2. Management notes that technical expertise of staff is more readily seen in their CVs and educational background than in the Human Resource database. Moreover, Bank teams work regularly with technical staff from other institutions, such as FAO.

3. Decentralization involved a large percentage of Bank staff being located in the field offices. The logic behind the decentralization was to delegate authority, functions, and staff to country offices weighing, on a case-by-case basis, the advantages of local responsiveness with the need to retain our global perspective—and balancing all this carefully against cost considerations (World Bank 2001b). More than 70 percent of country directors in the Africa Region today are based in client countries.

4. A long-standing research program of the World Bank Institute and the Research Department of the World Bank defines governance as the set of traditions and institutions for the exercise of authority in a country. The research reveals that the political, economic, and institutional dimensions of governance can be captured by six aggregate indicators: voice and accountability, political stability and absence of violence, government effectiveness (including the quality of public service, policy formulation, and government commitment), regulatory quality (including the ability of the government to formulate and implement sound policies and regulations), rule of law, and control of corruption (World Bank 2006i). While some of these aspects are discussed here, others are picked up in the section on policy and marketing reform in chapter 5.

5. Since the early 1990s, under the Special Program for African Agricultural Research (SPAAR), many national agricultural research systems (NARS) began rethinking their institutional model and moving away from top-down, supply-driven, publicly financed models toward more open and client-driven systems (CGIAR 2002).

Chapter 5

1. Management notes that agricultural technology projects support adoption of a range of technologies suitable for different agro-ecological conditions.

2. IEG's recent study of natural disaster assistance (IEG 2006c) also found that a many Bank projects can be characterized as ad hoc responses. The study also found that the Africa Region had the largest number of Bank-funded disaster projects and the lowest outcome rating.

3. Acute food insecurity results from short-term shocks (such as droughts) that reduce food availability, access, or utilization for an individual. Chronic food insecurity is limited access to food on a long-term basis and results from poverty, poor soil fertility, and food production and distribution systems with high unit costs.

4. A recent study (Anderson and others 2005) of links between CGIAR products and Bank operations, while noting important linkages, also notes that "the linkages and synergies between World Bank-financed projects, the IARC [International Agriculture Research Centers] research programs and the NARSs [National Agriculture Research Systems] in Eastern Africa have often been more by accident than by design" (p. 35).

5. A Country Assistance Evaluation for Malawi (IEG 2006e) that assessed the Bank's assistance to Malawi during fiscal 1996–2005 found that the Bank did not make an effective contribution to the development of the agriculture sector for various reasons:

> *First, the Bank's agriculture sector project implementation record has not been good. There have been six completed projects since FY96, only one of which has had a satisfactory outcome rating. Second, the Bank moved away from direct investments in the sector, addressing agricultural and rural economy issues primarily through multisector adjustment loans. This approach diluted the significance and impact of Bank interventions. Third, the Bank did very little policy analysis until 2003. Fourth, attempts to improve the quality of burley tobacco and strengthen farm-to-market links have not been successful. Finally, progress in developing Malawi's rural financial markets has been insubstantial. Thus, with respect to the sub objective for improved agricultural productivity and more efficient marketing, the outcome is unsatisfactory"* (IEG 2006e, p. 32).

It is worth noting that in the early 1990s, considerable policy analysis was carried out but it did not lead to effective results on the ground.

6. Only in fiscal 2006 did the Bank approve an irrigation, rural livelihood, and agriculture development project that is expected to contribute to increasing agriculture productivity.

7. Management notes that in the past two years, Malawi has had exceptional harvests and has exported maize. The success is not just because of good rains. It is also the result of a government program promoting access to inputs for smallholders, a fertilizer for work program under an IDA-financed Irrigation and Rural Livelihoods Project, and an overall improvement in the macro policy environment. Management also notes that the new generation of IDA-financed operations is contributing to food security and reducing the impact of drought by supporting irrigation and water harvesting, strengthening access to and supply of inputs, and supporting a number of critical institutional reforms and innovative approaches to risk management (including warehouse receipts, weather insurance, and commodity futures). IEG notes that 2004/05 had the lowest level of maize production since 1996–97. (See IEG's recent project assessment of the Malawi Emergency Drought Recovery Project, IEG 2007f) and that a number of the initiatives mentioned are still works in progress and it is too early to judge their impact.

8. IEG's natural disaster study found that Africa was also the only Region where borrowing for disasters was most often for droughts.

9. Management appreciates IEG's recognition of its role in supporting the development of improved, disease-resistant varieties of cassava. However, management would not see cassava as a missed opportunity for the Bank. The Bank does not specifically target production of cassava or other crops. Instead it supports countries in their efforts to improve the institutional setting for agriculture, generate technology improvements, and enhance the information available to producers to make their own decisions regarding production and marketing. In that context, reforms supported by the Bank that encouraged the removal of pan-territorial pricing for maize and fertilizer subsidies were key. These reforms resulted in a shift from maize to cassava in many marginal areas where cassava is the more suitable crop. The Bank's support for advisory services and technology dissemination, for example, under the Uganda National Agricultural Advisory Services Project, is the kind of activity that helps get farmers the information they need when they are deciding what crop to plant. (Management would also note in this context that the Bank serves as the implementing agency for IFAD's project on roots and tubers in Ghana.) Last, since cassava can be stored underground without harvesting for several years in the drier areas where it is produced, the countercyclical production with maize in figure 5.1 is probably largely increased harvesting instead of increased production per se.

10. In addition to the portfolio review, which was carried out on a sample of projects, this review looked at the objectives of all closed projects and found only one that included improving soil fertility as a project objective.

11. The initiative was launched in 1996 in response to concern from various stakeholders, with support from the World Bank, FAO, other donors, the CGIAR (represented by ICRAF), IFDC, and NGOs such as SG2000 and had as its original goal to help facilitate the introduction and adoption of sustainable soil fertility management practices by smallholder farmers.

12. The emphasis on environment also led to increase in natural resource management in Bank agricultural initiatives in the 1990s (World Bank 1991c). Other analytical work in the late 1990s (World Bank 1997a, p. 3) also noted that "most of the literature now agrees that the major environmental issue facing most of Africa is a combination of soil, water, forest, and pasture degradation in rural areas. The major cause is expansion of farming area resulting from growth of the rural population combined with farming practices that often mine the soils and cut forests for fuelwood and farming."

13. "In the late 1980s, sustainability emerged as a critical issue in African policy circles, because of famine, growing evidence of land degradation, deforestation and desertification and because of a rebirth of concern in developed countries for the environment. These forces translated into pressure on foreign assistance agencies to undertake environment programs, and in their interactions with African policy makers to insist on the urgency of addressing environmental problems" (Reardon 1998, p. 446).

14. The InterAcademy Council (2004, p. 202) report also notes, "A case can be made for selective subsidies on strategic inputs, such as fertilizers, until infrastructure can be improved to the extent that prices paid and received by African farmers are more in line with international competitors."

15. A recent study (Anderson and others 2005) found that, on average, well over one-half of the genetic material used for crop improvement in the East African

Community countries in commodities such as maize, cassava, beans, wheat, rice, legumes, and that involving agroforestry was provided directly by the CGIAR Centers to the national programs concerned.

16. The International Institute of Tropical Agriculture undertook a study to assess the level of adaptation and diffusion of the new extra-early maize technology since its introduction in villages in northern Nigeria in 1997. The study examined the rate of adoption of extra-early maize varieties and determined the factors influencing adoption and the constraints to adoption. Data were collected from 220 farming households in 14 villages. Out of 220 farmers selected in the random sample, only 20 farmers were growing the maize at the time of the survey. The major constraints to the adoption of extra-early maize varieties in the study villages were input related. Constraints cited very often by the adopters included unavailability of fertilizers (86.67 percent), unavailability of seed of extra-early maize (63.63 percent), labor constraints (36.67 percent), and land ownership problems (26.67 percent) (IITA 2004).

17. The dent hybrids [of maize] are much more vulnerable to damage by weevils in storage than the flinty local varieties. The introduction in the mid-1990s of the semi-flint varieties, which are more resistant to weevils, has made hybrids more popular than before. However, since they are not as resistant as local varieties, most farmers still prefer to grow both. While the breakthrough in breeding semi-flint hybrids has been important, the escalating cost of fertilizers and other inputs has made it difficult for farmers to grow more hybrid than local maize (Peters 2002).

18. Exposure to droughts and weather-related uncertainties affect farmers' incentives to adopt high-risk technologies and they often forgo available technologies that would require them to use fertilizers that would yield higher outputs, but are also riskier (Dercon and Christiaensen 2005).

19. "Public investment in marketing and transportation infrastructure would reduce input costs and increase producer prices by reducing transportation costs" (Ahmed, Sanders, and Nell 2000, p. 62).

20. Most African farmers currently know little about the range of plant varieties being developed and released by national crop improvement programs (Tripp and Rohrbach 2001). Effective smallholder seed supply systems are also still widely lacking in Africa (Sasakawa Africa Association 2004b).

21. Within a financial systems approach, financing for agriculture is seen as part of the wider rural finance market. In this approach, the appropriate role of the public sector is seen as ensuring that the environment is conducive to the emergence and growth of institutions adhering to commercial principles.

22. Supervision reports acknowledge that the rural and community banks lack the tools to successfully tackle agricultural lending.

23. An IEG review in 1996 found that the decline in Bank support for agricultural credit, which began in the early 1980s and persisted through 1992, continued in the middle 1990s. Another IEG study also noted the low lending for lines of credit in Africa over the period of fiscal 1993 and 2003 (IEG 2006h).

24. Management notes that the Bank has made a major commitment to increased transport in Africa, and virtually all of this benefits agriculture. Notably, passable trunk and secondary roads are a prerequisite for tertiary roads to play their role. Improvement has lagged in farm-to-market roads because of limited resources and the need to take on necessary priority investments first. In recent years, transport ministries in Africa supported by Bank teams typically work closely with agriculture ministries in setting transport priorities.

25. This section is an assessment of the overall extension approach, and not the subject areas where extension advice has been provided through Bank projects.

26. The InterAcademy Council sponsored a study to understand how to improve agriculture productivity and food security in Africa at the request of the then-Secretary General of the United Nations Kofi Annan.

27. "The Farmer Field School is a form of adult education, which evolved from the concept that farmers learn optimally from field observation and experimentation. It was developed to help farmers tailor their Integrated Pest Management (IPM) practices to diverse and dynamic ecological conditions." http://www.fao.org/docrep/006/ad487e/ad487e02.htm#P20_3691

28. The project assessment of the Tanzania Second Agricultural Extension Project notes that if public extension does wither and die, it seems likely that the poorer farmers and those predominantly producing food crops will suffer disproportionately. There is some global evidence that non-public extension, as might be expected, tends to target higher-income farmers. In Tanzania, whether such an approach could relieve sufficient budgetary burden at the top end to enable the

poor at the bottom end to be adequately covered is doubtful.

29. "The critiques of 'top-down' development and the call for more 'bottom-up' or participatory approaches should direct us not to oppose science/scientist to tradition/farmer but to help develop collaborative methods between rural producers and scientists/extension staff to identify, refine and circulate useful knowledge and 'best bets.' The aim is not to identify a single best solution for all times and places, but to recognize that multiple situations require multiple answers and that these necessarily change" (Peters 2002, p. 35).

30. "Pluralistic strategies often entail a change in roles and can run into active opposition of suspicious public agencies. In pursuing such a strategy, government requires a better understanding of existing extension services, and most cases suggested that the design of an extension policy supportive of a pluralistic system should begin with an inventory of the actors as in who provides what to whom, and an assessment of the quality of the services rendered before deciding on any reform" (World Bank 2004b, p. vii).

31. Christiaensen and others (2002), on the basis of their work in Africa, also found that households with larger private endowments such as land are in a better position to profit from new opportunities generated by liberalization and institutional change.

32. A value-cost ratio indicates the profitability of fertilizer application on crops.

33. That "getting prices right" could not—by itself—put African agriculture on the growth path was emphasized by the Bank-supported Managing Agriculture Development in Africa (MADIA) study way back in 1989, but evidently the lessons were not learned.

Dired cassava, Côte d'Ivoire. Photo by Ami Vitale, courtesy of the World Bank Photo Library.

An extensive bibliography that includes all sources consulted for this study can be found on the study Web site.

Ahmed, M. M., J. H. Sanders, and W. T. Nell. 2000. "New Sorghum and Millet Cultivar Introduction in Sub-Saharan Africa: Impacts and Research Agenda." *Agriculture Systems* 64: 55–65.

Anderson, Jock, Gershon Feder, and Sushma Ganguly. 2006. "The Rise and Fall of Training and Visit Extension: An Asian Mini-drama with an African Epilogue." Policy Research Working Paper 3928, World Bank, Washington, DC.

Anderson, Jock, M. Blackie, J. Kampen, I. Marzell, G. Semeho, and A. K. Seth. 2005. "Case Study of Links between CGIAR Products and Processes and World Bank Operations." Overview of a draft study as of November 2006. CGIAR Secretariat. World Bank, Washington, DC. (Available on request from janderson@worldbank.org.)

Ariga, Joshua, T. S. Jayne, and J. Nyoro. 2006. "Factors Driving the Growth in Fertilizer Consumption in Kenya, 1990–2005: Sustaining the Momentum in Kenya and Lessons for Broader Replicability in Sub-Saharan Africa." Tegemeo Working Paper 24, Egerton University, Nairobi.

Baffes, John. 2005. "The Cotton Problem." *The World Bank Research Observer* 20 (1): 109–44.

Bahigwa, Godfrey, Dan Rigby, and Philip Woodhouse. 2005. "Right Target, Wrong Mechanism? Agricultural Modernization and Poverty Reduction in Uganda." *World Development* 33(3): 481–96.

Bazaara, Nyangabyaki. 2001. "Impact of Liberalisation on Agriculture and Food Security in Uganda." Structural Adjustment Participatory Review Initiative (SAPRI), Uganda. Centre for Basic Research, Kampala.

Breman, Henk, and Siegfried Kofi Debrah. 2003. "Improving African Food Security." *SAIS Review* 23 (1): 153–70.

Camara, Oumou, and Ed Heinemann. 2006. "Overview of the Fertilizer Situation in Africa." Background Paper, African Fertilizer Summit, Abuja, Nigeria, June 9–13. http://www.africafertilizersummit.org/Background_Papers/index.html

CGAP (Consultative Group to Assist the Poorest). 2005. *Managing Risks and Designing Products for Agricultural Microfinance: Features of an Emerging Model.* Occasional Paper No. 11. Washington, DC: World Bank.

CGIAR (Consultative Group for International Agricultural Research). 2002. "Annual General Meeting 2002. Stakeholder Meeting Agenda Item 6. CGIAR and Sustainable Development: 2002 Summits and the Way Ahead." http://www.cgiar.org/meetings/agm02/index.html

Chapman, Robert, and Robert Tripp. 2003. "Changing Incentives for Agricultural Extension—A Review of Privatized Extension in Practice." ODI Agricultural Research and Extension Network Paper No. 12. London: Overseas Development Institute.

Christiaensen, Luc, and Lionel Demery. 2007. *Down to Earth—Agriculture and Poverty Reduction in Africa*. Directions in Development Series. Washington, DC: World Bank.

Christiaensen, Luc, Lionel Demery, and Stefao Paternostro. 2002. *Growth, Distribution, and Poverty in Africa: Messages from the 1990s*. Washington, DC: World Bank.

Collier, Paul, and Jan Willem Gunning. 1997. "Explaining African Economic Performance." WPS97-2.1, Center for the Study of African Economics, Oxford, U.K.

Commission for Africa. 2005. *Our Common Interest*. Glasgow: DFID.

DAC (Development Assistance Committee). 2007. "The World Bank Group's Africa Action Plan: Progress in Implementation." DC2007–0010. World Bank and International Monetary Fund, Washington, DC.

Dercon, Stefan, and Luc Christiaensen. 2005. "Consumption Risk, Technology Adoption and Poverty Traps: Evidence from Ethiopia." Policy Research Working Paper 4257, World Bank, Washington, DC.

DFID (U.K. Department for International Development). 2004. "Official Development Assistance to Agriculture." http://dfid-agriculture-consultation.nri.org/summaries/wp9.pdf

Diao, Xinshen, Peter Hazell, Danielle Resnick, and James Thurlow. 2006. "The Role of Agriculture in Development: Implications for Sub-Saharan Africa." DGSD Discussion Paper No. 29, International Food Policy Research Institute, Washington DC.

Dixon, J., A. Gulliver, and D. Gibbon. 2001. *Farming Systems and Poverty: Improving Farmers' Livelihood in a Changing World.* Rome and Washington, DC: FAO and the World Bank.

Donovan, Graeme, and Frank Casey. 1998. "Soil Fertility Management in Sub-Saharan Africa." Technical Paper No. 408, World Bank, Washington, DC.

Dorosh, Paul, and Steven Haggblade. 2003. "Growth Linkages, Price Effect, and Income Distribution in Sub-Saharan Africa." *Journal of African Economies* 12(2): 207–35.

Doss, Cheryl R. 1999. *Twenty-Five Years of Research on Women Farmers in Africa: Lessons and Implications for Agricultural Research Institutions with an Annotated Bibliography.* Economic Program Paper 99-02. Mexico, DF, Mexico: CIMMYT.

ECA (Economic Commission for Africa). 2003. "Towards a Green Revolution in Africa: Harnessing Science and Technology for Sustainable Modernization of Agriculture and Rural Transformation." Sustainable Development Division, ECA, Addis Ababa, Ethiopia.

Ehui, Simeon, and John Pender. 2005. "Resource Degradation, Low Productivity, and Poverty in Sub-Saharan Africa: Pathways Out of the Spiral." *Agricultural Economics* 32(1).

Eicher, Carl K. 1999. "Institutions and the African Farmer." Third Distinguished Economist Lecture, Mexico City, CIMMYT Economics Program, January 22. *Issues in Agriculture* 14. Mexico D.F., Mexico.

Eicher, Carl K., and Mandivamba Rukuni. 2003. "The CGIAR in Africa: Past, Present, and Future." Thematic Working Paper for *The CGIAR at 31: An Independent Meta-Evaluation of the Consultative Group on International Agricultural Research.* Independent Evaluation Group. Washington, DC: World Bank.

Eilitta, Marjatta. 2006. "Achieving an African Green Revolution: A Vision for Sustainable Agricultural Growth in Africa." Background Paper, African Fertilizer Summit, Abuja, Nigeria, June 9–13. http://www.africafertilizersummit.org/Background_Papers/index.html

Ellis, Frank, and Godfrey Bahigwa. 2003. "Livelihood and Rural Poverty Reduction in Uganda." *World Development* 31(6): 997–1013.

Essama-Nssah, B., and James J. Gockowski. 2000. *Cameroon Forest Sector: Development in a Difficult Political Economy*. Evaluation Country Case Study Series. Washington, DC: World Bank.

Evenson, Robert E. 2003. "Production Impacts of Crop Genetic Improvement." In *Crop Variety Improvement and Its Effect on Productivity: The Impact of International Agricultural Research*, ed. R. E. Evenson and D. Gollin. Wallingford, Oxon, U.K.: CABI.

Evenson, Robert E., and D. Gollin, eds. 2003. *Crop Variety Improvement and Its Effect on Productivity: The Impact of International Agricultural Research.* Wallingford, Oxon, U.K.: CABI.

FAO (Food and Agriculture Organization). 2001. *Global Farming Systems Study: Challenges and Priorities to 2030. Synthesis and Global Overview.* Rome: FAO.

FAO and IFAD (International Fund for Agricultural Development). 2005. "A Review of Cassava in Africa." In *Proceedings of the Validation Forum on the Global Cassava Development Strategy*, Volume 2. Rome: FAO.

Foster, Mick, Adrienne Brown, and Felix Naschold. 2001. "Sector Programme Approaches: Will They Work in Agriculture?" *Development Policy Review* 19 (3).

Gladwin, Christina. 2002. "Gender and Soil Fertility in Africa: An Introduction." *African Studies Quarterly* 6(1/2) [online] URL:http://web.africa.ufl.edu/asq/v6/v6i1a1.htm.

Hartmann, P. 2004. *An Approach to Hunger and Poverty Reduction for Sub-Saharan Africa.* Ibadan, Nigeria: International Institute of Tropical Agriculture.

Hazell, Peter, and Joachim von Braun. 2006. "Aid to Agriculture, Growth, and Poverty Reduction." *EuroChoices* 5(1): 6–13.

Heerink, Nico. 2005. "Soil Fertility Decline and Economic Policy Reform in Sub-Saharan Africa." *Land Use Policy* 22: 67–74.

Henao, Julio, and Carlos Baanante. 2006. "Agricultural Production and Soil Nutrient Mining in Africa: Implications for Resource Conservation and Policy Development." Background Paper, African Fertilizer Summit, Abuja, Nigeria, June 9–13. http://www.africa fertilizersummit.org/Background_Papers/ index.html

Hine, John, and John Rutter. 2000. "Roads, Personal Mobility and Poverty: The Challenge." Paper for the Transport and Poverty Alleviation Workshop, Wokingham, U.K., Transport Research Laboratory.

Howard, Julie, Valerie Kelly, Julie Stepanek, Eric W Crawford, Mulat Demeke, and Mywish Mredia. 1999. "Green Revolution Technology Takes Root in Africa." International Development Working Paper No. 76, Department of Agricultural Economics, Michigan State University, East Lansing.

IEG (Independent Evaluation Group). 2007a. "Ethiopia: National Fertilizer Sector Project." Project Performance Assessment Report, World Bank, Washington, DC.

———. 2007b. "Ethiopia: Seed System Development Project." Project Performance Assessment Report, World Bank, Washington, DC.

———. 2007c. "Madagascar: Agricultural Extension Program Support." Project Performance Assessment Report, World Bank, Washington, DC.

———. 2007d. "Madagascar: Irrigation Project." Project Performance Assessment Report, World Bank, Washington, DC.

———. 2007e. "Madagascar: Second Irrigation Rehabilitation Project." Project Performance Assessment Report, Washington, DC: World Bank.

———. 2007f. "Malawi: Emergency Drought Recovery Project." Project Performance Assessment Report, World Bank, Washington, DC.

———. 2007g. "Mali: Agricultural Trading And Processing Promotion Pilot Project." Project Performance Assessment Report, World Bank, Washington, DC.

———. 2007h. "Mali: National Agricultural Research Project." Project Performance Assessment Report, World Bank, Washington, DC.

———. 2007i. "Mali: Pilot Private Irrigation Promotion Project." Project Performance Assessment Report, World Bank, Washington, DC.

———. 2007j. "Tanzania: Agricultural Research Project 2." Project Performance Assessment Report, World Bank, Washington, DC.

———. 2007k. "Tanzania: Agricultural Sector Management Project." Project Performance Assessment Report, World Bank, Washington, DC.

———. 2007l. "Tanzania: National Extension Project 2." Project Performance Assessment Report, World Bank, Washington, DC.

———. 2007m. "Zambia: Emergency Drought Recovery Project." Project Performance Assessment Report, World Bank, Washington, DC.

———. 2007n. *The Development Potential of Regional Programs: An Evaluation of World Bank Support for Multicountry Operations.* IEG Study Series. Washington, DC: World Bank.

———. 2007o. *A Decade of Action in Transport: An Evaluation of World Bank Assistance to the Transport Sector, 1995–2005.* IEG Study Series. Washington, DC: World Bank.

———. 2006a. *Assessing World Bank Support for Trade, 1987–2004. An IEG Evaluation.* IEG Study Series. Washington, DC: World Bank.

———. 2006b. *Engaging with Fragile States: An IEG Review of World Bank Support to Low-Income Countries Under Stress.* IEG Study Series. Washington, DC: World Bank.

———. 2006c. *Hazards of Nature, Risks to Development: An IEG Evaluation of World Bank Assistance for Natural Disasters.* IEG Study Series. Washington, DC: World Bank.

———. 2006d. *Madagascar Country Assistance Evaluation.* Washington, DC: World Bank.

———. 2006e. *Malawi Country Assistance Evaluation.* Washington, DC: World Bank.

———. 2006f. *Senegal Country Assistance Evaluation.* Report No. 36286. Washington, DC: World Bank.

———. 2006g. *The World Bank's Assistance to Agricultural Water Management (1994–2004): An IEG Portfolio Review.* IEG Study Series. Washington, DC: World Bank.

———. 2006h. *World Bank Lending for Lines of Credit: An IEG Evaluation.* IEG Study Series. Washington, DC: World Bank.

———. 2006i. *Annual Review of Development Effectiveness 2006: Getting Results.* IEG Study Series. Washington, DC: World Bank.

———. 2005a. *The Effectiveness of World Bank Support for Community-Based and -Driven Development.* IEG Study Series. Washington, DC: World Bank.

———. 2005b. "Kenya Arid Lands Resource Management Project." Project Performance Assessment Report, Report No. 34052, World Bank, Washington, DC.

———. 2004a. *Rwanda Country Assistance Evaluation.* Report No. 27568. Washington, DC: World Bank.

———. 2004b. *2003 Annual Review of Development Effectiveness: The Effectiveness of Bank Support for Pol-*

icy Reform. IEG Study Series. Washington, DC: World Bank.

———. 2003a. "Republic of Benin: Agricultural Services Restructuring Project, Community-Based Food Security Project, Second Rural Savings and Loans Cooperatives Rehabilitation Project." Project Performance Assessment Report, Report No. 26207, World Bank, Washington, DC.

———. 2003b. *The Poverty Reduction Strategy Initiative—An Independent Evaluation of the World Bank's Support Through 2003.* IEG Study Series. Washington, DC: World Bank.

———. 2003c. "Zambia: Agriculture Sector Investment Program." Project Performance Assessment Report, World Bank, Washington, DC.

———. 2003d. "Zimbabwe: Second Structural Adjustment Credit." Project Performance Assessment Report, Report No. 25479, World Bank, Washington, DC.

———. 2003e. "Mali: Natural Resources Management Project." Project Performance Assessment Report, World Bank, Washington, DC.

———. 2002a. "Mozambique: Project Performance Assessment." Report No. 23630, World Bank, Washington, DC.

———. 2002b. *Toward Sharpening the Focus on Rural Poverty: A Review of World Bank Experience.* Washington, DC: World Bank.

———. 2002c. *Lesotho: Development in a Challenging Environment. A Joint World Bank–African Development Bank Evaluation.* IEG Study Series. Washington, DC: World Bank.

———. 2001. "Ghana: Performance Assessment Reports and Sectoral Overview." World Bank, Washington, DC.

———. 2000a. *Rural Development—From Vision to Action? (Phase II).* Washington, DC: World Bank.

———. 2000b. *Kenya Country Assistance Evaluation.* Washington, DC: World Bank.

———. 1999a. *Rural Development: From Vision to Action?* Washington, DC: World Bank.

———. 1999b. "Uganda: Agricultural Sector Adjustment Credit." Performance Audit Report, World Bank, Washington, DC.

———. 1999c. "Capacity Building in the Agriculture Sector in Africa." *Précis* Number 180. Washington, DC: IEG.

———. 1998a. *The Special Program of Assistance for Africa (SPA): An Independent Evaluation.* Washington, DC: World Bank.

———. 1998b. "Zambia: First and Second Privatization and Industrial Reform Credits and Economic and Social Adjustment Credit." Performance Audit Report, World Bank, Washington, DC.

———. 1998c. *The World Bank and the Agricultural Sector in Kenya: An OED Review.* IEG Study Series. Washington, DC: World Bank.

———. 1998d. *Malawi Impact Evaluation Report: The World Bank and the Agriculture Sector.* Washington, DC: World Bank.

———. 1997a. *Zambia Country Assistance Review: Turning an Economy Around.* Washington, DC: World Bank.

———. 1997b. *Agricultural Extension and Research Achievements and Problems in National Systems: A World Bank Operations Evaluation Study.* Washington, DC: World Bank.

———. 1996. *A Review of Bank Lending for Agricultural Credit and Rural Finance (1998–1992).* Washington, DC: World Bank.

———. 1995. *Structural and Sectoral Adjustment: World Bank Experience, 1980–92.* Washington, DC: World Bank.

———. 1994. "Madagascar: Agricultural Sector Adjustment Credit." Project Audit Report. No. 13081, World Bank, Washington, DC.

IFDC (International Center for Soil Fertility and Agricultural Development). 2006. "Input Subsidies and Agricultural Development: Issues and Options for Developing and Transitional Economies." Background Paper, African Fertilizer Summit, Abuja, Nigeria, June 9–13. http://www.africafertilizersummit.org/Background_Papers/index.html

IFPRI (International Food Policy Research Institute). 2006a. *The Role of Agriculture in Development: Implications for Sub-Saharan Africa.* Development Strategy and Governance Division (DSGD) Discussion Paper No. 29. Washington, DC: IFPRI.

———. 2006b. "Are Farmer Field Schools the Answer to Extension Problems?" *IFPRI Forum*, September: 3.

———. 2005a. *Looking Ahead: Long Term Prospects for Africa's Agricultural Development and Food Security.* 2020 Discussion Paper 41. Washington, DC: IFPRI.

———. 2005b. *Achieving Regional Growth Dynamics in African Agriculture.* Development Strategy and Governance Division (DSGD) Discussion Paper No. 17. Washington, DC: IFPRI.

———. 2004a. "African Agriculture: Past Performance,

Future Imperatives." In *2020 Focus No. 12: Building on Successes in African Agriculture,* ed. Steven Haggblade, Brief 1. Washington, DC: IFPRI.

———. 2004b. "Assuring Food and Nutrition Security in Africa by 2020." Proceedings of an All-Africa Conference, Kampala, Uganda, April 1-3 2004. IFPRI, Washington, DC.

———. 2002a. *Governance and Food Security in an Age of Globalization*. Food, Agriculture, and the Environment Discussion Paper 36. Washington, DC: IFPRI.

———. 2002b. *Ending Hunger in Africa: Only the Small Farmer Can Do It.* Washington, DC: IFPRI.

———. 2000. *The Road Half Traveled: Agricultural Market Reform in Sub-Saharan Africa.* Washington, DC: IFPRI.

———. 1993. "Aid to Agriculture: Reversing the Decline in Food." Policy Statement No. 17. Washington, DC: IFPRI.

IITA (International Institute of Tropical Agriculture). 2004. *Program for Improving and Intensifying Cereal Legume Systems in the Moist and Dry Savannas of West and Central Africa.* Annual Report 2004. Croyden, U.K.: IITA.

Institute of Development Studies. 2006. "Achieving Food Security: What Next for Sub-Saharan Africa?" *ID21 Insights* 61: 1–2.

InterAcademy Council. 2004. *Realizing the Promise and Potential of African Agriculture.* Amsterdam: InterAcademy Council.

Jayne, T. S., and Stephen Jones. 1997. "Food Marketing and Pricing Policy in Eastern and Southern Africa: A Survey." *World Development* 25(9): 1505–27.

——. 1996. "Food Marketing and Pricing Policy in Eastern and Southern Africa: Lessons for Increasing Agricultural Productivity and Access to Food." Policy Synthesis for USAID Bureau for Africa Office of Sustainable Development, No. 12.

Jayne, T. S., James D. Shaffer, John M. Staatz, and Thomas Reardon. 1997. *Improving the Impact of Market Reform on Agricultural Productivity in Africa: How Institutional Design Makes a Difference.* MSU International Development Working Papers No. 66, Department of Agricultural Economics, Michigan State University, East Lansing.

Jayne, T. S., Mulinge Mukumbu, Munhamo Chisvo, David Tschirley, Ballard Zulu, Michael T. Weber, Robert Johansson, Paula Santos, and David Soroko. 1999. "Successes and Challenges of Food Market Reform Experiences from Kenya, Mozambique, Zambia and Zimbabwe." International Development Working Paper Series No. 72, Department of Agricultural Economics, Michigan State University, East Lansing.

Johnson, Alison, Matthew Martin, and Hannah Bargawi. 2004. "The Effectiveness of Aid to Africa since the HIPC Initiative: Issues, Evidence and Possible Areas for Action." Background Paper for Commission for Africa, Development Finance International, Bethesda, MD. Draft.

Kane, Sam, and Carl K. Eicher. 2004. "Foreign Aid and the African Farmer." Staff Paper 2004-13, Department of Agricultural Economics, Michigan State University, East Lansing.

Kelly, Valerie, Akinwumi A. Adesina, and Ann Gordon. 2003. "Expanding Access to Agricultural Inputs in Africa: A Review of Recent Market Development Experience." *Food Policy* 28: 379–404.

Kelly, Valerie A., Eric W. Crawford, and T. S. Jayne. 2003. "Agricultural Input Use and Market Development in Africa: Recent Perspectives and Insights." Policy Synthesis 70, Department of Agricultural Economics, Michigan State University, East Lansing.

Kherallah, Mylene, Christopher Delgado, Eleni Gabre-Madhin, Nicholas Minot, and Michael Johnson. 2002. *Reforming Agricultural Markets in Africa.* Washington, DC: IFPRI.

Kritzinger-van Niekerk, L., and Fabrice Houdart. 2005. "New Partnership for Africa's Development." September. http://siteresources.worldbank.org/INTAFR REGINICOO/Resources/NEPAD_Brief.pdf

Lipton, Michael, Julie Litchfield, and Jean-Marc Faures. 2003. "The Effects of Irrigation on Poverty: A Framework for Analysis." *Water Policy* 5: 413–27.

Maredia, Mywish K., and David A. Raitzer. 2006. *CGIAR and NARS Partner Research in Africa. Evidence of Impact to Date.* Rome: CGIAR Science Council Secretariat.

Maumbe, B., and M. Swinton. 2003. "Hidden Health Costs of Pesticide Use in Zimbabwe's Smallholder Cotton Growers." *Social Science and Medicine* 57 (9): 1559–71.

Mekuria, Mulugetta, and Stephen R. Waddington. 2002. "Initiatives to Encourage Farmer Adoption of Soil-Fertility Technologies for Maize-Based Cropping Systems in Southern Africa." In *African Agriculture: Understanding and Improving Current Practices,* ed. C. Barrett, F. Place, and A. A. Aboud. Wallingford, U.K., and New York, NY: CABI Publishing in association with the International Center for Research in Agroforestry.

Mellor, John W. 1998. "Foreign Aid and Agriculture-Led Development." In *International Agricultural Development,* ed. Carl K Eicher and John Staatz , 55-67. 3rd ed. Baltimore: Johns Hopkins University Press.

Muyanga, Milu, and T. S. Jayne. 2006. "Agricultural Extension in Kenya: Practice and Policy Lessons." Tegemeo Institute of Agricultural Policy and Development Working Paper 21, Edgerton University, Nairobi.

NBER (National Bureau of Economic Research). 2002. "When Economic Reform Goes Wrong: Cashews in Mozambique." Working Paper 9117, NBER, Cambridge, MA.

Ndiritu, Cyrus G., John K. Lynam, and Adiel N. Mbabu, eds. 2004. *Transformation of Agricultural Research Systems in Africa: Lessons From Kenya.* East Lansing: Michigan State University Press.

NEPAD (New Partnership for Africa's Development). 2004. "NEPAD Targets Cassava as Africa's Top Fighter against Poverty." NEPAD Dialogue: Focus on Africa 36. http://www.un.org/special-rep/ohrlls/News_ flash 2004/NEPAD%20Newsletter%20English% 2036.htm

Nweke, Felix. 2004. "New Challenges in the Cassava Revolution in Nigeria and Ghana." EPTD Discussion Paper No. 118, Washington, DC, IFPRI.

Nyoro, James, Lilian Kirimi, and T. S. Jayne. Undated. *Competitiveness of Kenyan and Ugandan Maize Production: Challenges for the Future.* Tegemeo Institute of Agricultural Policy and Development. Nairobi: Egerton University.

OECD (Organisation for Economic Co-operation and Development). 2001. "Aid to Agriculture." http://www .oecd.org /dataoecd/40/43/2094403.pdf.

OED (Operations Evaluation Department) Renamed Independent Evaluation Group in December 2005. Please see IEG for publications.

Orr, Alastair. 2000. "Green Gold? Burley Tobacco, Smallholder Agriculture, and Poverty Alleviation in Malawi." *World Development* 28(2): 347–63.

Palmer, Keith. 2004. "Achieving Higher Growth and Poverty Reduction in Sub-Saharan Africa: A Note for the Commission on Africa." Cambridge Economic Policy Associates. http://www.keithpalmer.org/pdf2.pdf

Peacock, Tony, Christopher Ward, and Gretel Gambarelli. 2007. "Investment in Agricultural Water for Poverty Reduction and Economic Growth in Sub-Saharan Africa Synthesis Report" (draft). Collaborative program of ADB, FAO, IFAD, IWMI, and World Bank.

Peters, Pauline. 2002. "The Limits of Knowledge: Securing Rural Livelihoods in a Situation of Resource Scarcity." In *Natural Resources Management— African Agriculture: Understanding and Improving Current Practices,* ed. C. Barrett, F. Place, and A. A. Aboud. Wallingford, U.K., and New York, NY: CABI Publishing in association with the International Center for Research in Agroforestry.

Poulton, Colin, Jonathan Kydd, and Andrew Dorward. 2006. "Overcoming Market Constraint on Pro-Poor Agricultural Growth in Sub-Saharan Africa." *Development Policy Review* 24: 243–77.

QAG (Quality Assurance Group, The World Bank). 2006. *2006 Annual Review of Portfolio Performance.* Washington, DC: QAG.

———. 2005. "Quality of Country AAA: A QAG Assessment." QAG, Washington, DC.

———. 2004. "Cross-Country ARD Sector AAA Assessment." QAG, Washington, DC.

Reardon, Thomas. 1998. "African Agriculture: Productivity and Sustainability Issues." In *International Agricultural Development*, ed. Carl K. Eicher and John Staatz. 3rd ed. Baltimore: Johns Hopkins University Press.

Sanchez, Pedro A. 2002. "Soil Fertility and Hunger in Africa." *Science* 295: 2019–20.

Sanchez, P. A., K. D. Shepherd, M. J. Soule, F. M. Place, R. J. Buresh, A. N. Izac, A. U. Mokwunye, F. R. Kwesiga, C. G. Ndiritu, and P. L. Woomer. 1997. "Soil Fertility Replenishment in Africa: An Investment in Natural Resource Capital." In *Replenishing Soil Fertility in Africa*, ed. R. J. Buresh, P. A. Sanchez, and F. G. Calhoun. Madison, WI: Soil Science Society of America.

Sanders, John H., Barry I. Shapiro, and Sunder Ramaswamy. 1996. *The Economics of Agricultural Technology in Semiarid Sub-Saharan Africa.* Baltimore: John Hopkins University Press.

Sasakawa Africa Association. 2006. "Green to Gene Revolution: Feeding the Future." *Newsletter* Issue 22. http://www.saa-tokyo.org/english/

———. 2005. *Newsletter* Issue 21. http://www.saa-tokyo.org/english/

———. 2004a. *Annual Report 2003-2004.* http://www .saa-tokyo.org/english/

———. 2004b. *Newsletter* Issue 20. http://www.saa-tokyo.org/english/

———. 2001. *Newsletter* Issue 16. http://www.saa-tokyo.org/english/

Scoones, Ian, ed. 2001. *Dynamics and Diversity: Soil Fertility and Farming Livelihoods in Africa*. London and Sterling VA: Earthscan.

Shapouri, S., and S. L. Rosen. 2001. "Toll on Agriculture from HIV/AIDS in Sub-Saharan Africa." *Agriculture Information Bulletin* 765–9.

SIDA (Swedish International Development Cooperation Agency). 2006. *Addressing Food Crisis in Africa*. SIDA Division for Rural Development. http://www.sida.se/sida/jsp/sida.jsp?d=118&a=19308&language=en_US

Toulmin, Camilla. 2006. *Securing Land and Property Rights in Sub-Saharan Africa: The Role of Local Institutions.* (Revised version of *Securing Land and Property Rights in Africa: Improving the Investment Climate. Global Competitiveness Report 2005–06.*) World Economic Forum, Switzerland.

Tripp, Robert, and David Rohrbach. 2001. "Policies for African Seed Enterprise Development." *Food Policy* 26: 147–61.

Tschirley David, Colin Poulton, and Duncan Boughton. 2006a. *The Many Paths of Cotton Sector Reform in Eastern and Southern Africa: Lessons from a Decade of Experience.* International Development Working Papers, Department of Agricultural Economics, Michigan State University, East Lansing.

Tschirley, David, Jan J. Nijhoff, Pedro Arlindo, Billy Mwinga, Michael T. Weber, and T. S. Jayne. 2006b. "Anticipating and Responding to Drought Emergencies in Southern Africa: Lessons from the 2002–2003 Experience." International Development Working Papers, Department of Agricultural Economics, Michigan State University, East Lansing.

Tshibaka, Tshikala B. 2003. "Economic Policy Reforms, External Factors, and Domestic Agricultural Terms of Trade in Selected West African Countries." In *African Voices on Structural Adjustment,* ed. Thandika Mkandawire and Charles C. Soludo. Ottawa: The International Development Research Centre.

UNDP (United Nations Development Programme). 2006. *Land Rights Reform and Governance in Africa: How to Make it Work in the 21st Century?* Discussion Paper. New York: UNDP.

———. 2005. *Halving Hunger: It Can Be Done*. UN Millenium Project Task Force on Hunger. London and Sterling, VA: Earthscan.

van den Brink, Rogier, Glen Thomas, Hans Binswanger, John Bruce, and Frank Byamugisha. 2005. "Consensus, Confusion, and Controversy: Selected Land Reform Issues in Sub-Saharan Africa." Working Paper No. 71, World Bank, Washington, DC.

Whingwiri, E. 2004. "NEPAD Pan Africa Cassava Initiative." Presented at ISTRC-AB Symposium, 31 October–5 November, Mombassa, Kenya.

Whyte, M. A., and D. Kyaddondo. 2006. "We Are Not Eating Our Own Food Here: Food Security and the Cash Economy in Eastern Uganda." *Land Degradation and Development* 17: 173–82.

Wiggins, Steve. 2005. "Success Stories from African Agriculture: What Are the Key Elements of Success?" *IDS Bulletin* 36(2).

———. 2000. "Interpreting Changes from the 1970s to the 1990s in African Agriculture through Village Studies." *World Development* 28(4): 631–62.

World Bank. 2007a. *Fertilizer Use in African Agriculture: Lessons Learned and Good Practice Guidelines*. Directions in Development Series. Washington, DC: World Bank.

———. 2007b. *Reforming Agricultural Trade for Developing Countries.* Vols. 1 and 2. Washington, DC: World Bank.

———. 2007c. *World Development Report 2008: Agriculture for Development*. Washington, DC: World Bank.

———. 2007d. "Tanzania— Pilot Rural Investment Climate Assessment: Stimulating Non-farm Microenterprise Growth" (draft). World Bank, Washington, DC.

———. 2007e. "Zambia—Smallholder Agricultural Commercialization Strategy." World Bank, Washington, DC.

———. 2006a. *Reengaging in Agricultural Water Management: Challenges and Options.* Directions in Development Series. Washington, DC: World Bank.

———. 2006b. *Alternative Approaches for Promoting Fertilizer Use in Africa Agriculture and Rural Development.* Discussion Paper 22. Washington, DC: World Bank.

———. 2006c. *Factors Affecting Demand for Fertilizer in Sub-Saharan Africa.* Agriculture and Rural Development Discussion Paper 23. Washington, DC: World Bank.

———. 2006d. *Increasing Fertilizer Use in Africa: What Have We Learned?* Agriculture and Rural Development Discussion Paper 25. Washington, DC: World Bank.

———. 2006e. "Development of the Cotton Sector in West and Central Africa." Presentation by Gobind Nankani, Vice President of the Africa Region, World

Bank, Wilson Center, Conference on "Cotton—The Next Steps," October 26. World Bank, Washington, DC.

———. 2006f. "Strategies for Cotton in West and Central Africa: Enhancing Competitiveness in the 'Cotton 4.'" Working Paper 108, World Bank, Washington, DC.

———. 2006g. *Market Led Smallholder Development in the Zambezi Valley*. Report No. 35466. Washington, DC: World Bank.

———. 2006h. *Ethiopia: Managing Water Resources to Maximize Sustainable Growth.* Country Water Resources Assistance Strategy. Washington, DC: World Bank.

———. 2006i. *A Decade of Measuring the Quality of Governance—Governance Matters 2006: Worldwide Governance Indicators.* Washington, DC: World Bank.

———. 2006j. "Promoting Increased Fertilizer Use in Africa: Lessons Learned and Good Practice Guidelines." Africa Fertilizer Strategy Assessment. ESW Technical Report. Discussion draft. World Bank Papers. Background Paper, African Fertilizer Summit, Abuja, Nigeria, June 9–13. http://www.africafertilizer summit.org/Background_Papers/index.html

———. 2006k. "Factors Affecting Supply of Fertilizer in Sub-Saharan Africa." World Bank Papers. Background Paper, African Fertilizer Summit, Abuja, Nigeria, June 9–13. http://www.africafertilizersummit.org/Back ground_Papers/index.html

———. 2006l. "Project Appraisal Document on A Proposed Credit in the Amount of SDR 61.6 Million to the United Republic of Tanzania for an Agricultural Sector Development Project." Africa Region, World Bank, Washington, DC.

———. 2006m. "Project Appraisal Document on a Proposed Grant in the Amount of SDR 25.7 Million to the Republic of Zambia for an Agricultural Development Support Project." Africa Region, World Bank, Washington, DC.

———. 2005a. *Meeting Development Challenges: Renewed Approaches to Rural Finance.* Washington, DC: World Bank.

———. 2005b. *Uganda: Policy Options for Increasing Crop Productivity and Reducing Soil Nutrient Depletion and Poverty.* Environment and Socially Sustainable Development, Africa Region. Washington, DC: World Bank.

———. 2005c. *Agriculture Investment Sourcebook.* Washington, DC: World Bank.

———. 2005d. *Rural Finance Innovations: Topics and Case Studies.* Washington, DC: World Bank.

———. 2005e. *Meeting the Challenge of Africa's Development: A World Bank Group Action Plan.* Africa Region. Washington, DC: World Bank.

———. 2005f. *Pro-Poor Growth in the 1990s: Lessons and Insights from 14 Countries. Operationalizing Pro-Poor Growth Research Program*. Washington, DC: World Bank.

———. 2005g. *Conversations with Country Directors on the Agriculture and Rural Development Agenda*. Washington, DC: World Bank.

———. 2005h. *Food Aid and Food Security in the Short and Long Run: Country Experience from Asia and Sub-Saharan Africa*. SP Discussion Paper No. 0538. Washington, DC: World Bank.

———. 2005i. *Agricultural Growth for the Poor: An Agenda for Development.* Washington, DC: World Bank.

———. 2005j. *Burkina Faso: Reducing Poverty Through Sustained Equitable Growth*. Report No. 29743. Washington, DC: World Bank.

———. 2005k. *Agriculture, Rural Development and Pro-Poor Growth: Country Experiences in the Post-Reform Era.* Agriculture and Rural Development Discussion Paper 21. Washington, DC: World Bank.

———. 2004a. *Strategic Framework for Assistance to Africa: IDA and the Emerging Partnership Model.* Washington, DC: World Bank.

———. 2004b. "Demand-Driven Approaches to Agriculture Extension: Case Studies of International Initiatives," ed. William Rivera and Gary Alex, Working Paper 31890, Washington, DC, World Bank.

———. 2004c. *Operationalising Pro-Poor Growth: A Country Case Study on Burkina Faso.* A Joint Initiative of AFD, BMZ (GTZ, KfW Development Bank), DFID and the World Bank. Washington, DC: World Bank.

———. 2004d. "ALive: A Partnership for Livestock Development for Poverty Reduction and Economic Growth in Sub-Saharan Africa." Draft Concept Note, World Bank Economic and Social Sustainable Development Department, Africa Region, Washington, DC. alive-online.org/data/files/ALive_CN_(Nov 04).doc

———. 2004e. *The Cotton Problem*. Washington, DC: World Bank.

———. 2003a. *Land Policies for Growth and Poverty*

Reduction. Policy Research Report. Washington, DC: World Bank.

———. 2003b. *Commodity Market Reform in Africa: Some Recent Experiences.* Policy Research Working Paper 2995. Development Research Group. Washington, DC: World Bank.

———. 2003c. "FY 2003 Agriculture and Rural Portfolio Review." ARD Working Paper, World Bank, Washington, DC.

———. 2003d. *Reaching the Rural Poor: A Renewed Strategy for Rural Development*. Washington, DC: World Bank.

———. 2003e. "Country Assistance Strategy for the Federal Democratic Republic of Ethiopia." Africa Region, World Bank, Washington, DC.

———. 2003f. "Togo—National Agricultural Services Support Project." Africa Region, World Bank, Washington, DC.

———. 2002a. *From Action to Impact: The Africa Region's Rural Strategy.* Washington, DC: World Bank.

———. 2002b. "Rural Extension and Advisory Services: New Directions." Rural Strategy Background Paper No. 9. World Bank, Washington, DC.

———. 2002c. *FY2001 Rural Portfolio Review*. Washington, DC: World Bank.

———. 2002d. *Reinvesting in African Small-Holder Agriculture: The Role of Tree Crops in Sustainable Farming Systems.* Washington, DC: World Bank.

———. 2001a. *Adjustment Lending Retrospective.* Operations Policy and Country Services. Washington, DC: World Bank.

———. 2001b. *World Bank Strategic Framework.* Washington, DC: World Bank.

———. 2001c. *Livestock Development: Implications for Rural Poverty, the Environment and Global Food Security.* Directions in Development Series. Washington, DC: World Bank.

———. 2000. *Can Africa Claim the 21st Century?* Washington, DC: World Bank.

———. 1998. *Evolution of Agricultural Services in Sub-Saharan Africa: Trends and Prospects*. World Bank Discussion Paper 390. Africa Region Series. Washington, DC: World Bank.

———. 1997a. *Rural Development Strategies for Poverty Reduction and Environmental Protection in Sub-Saharan Africa*. Directions in Development. Washington, DC: World Bank.

——— 1997b. *Rural Finance: Issues, Design and Best Practice*. ESSD Studies and Monographs Series 14. Washington, DC: World Bank

———. 1997c. *Rural Development: From Vision to Action: A Sector Strategy.* Washington, DC: World Bank.

———. 1994. *Adjustment in Africa: Reforms, Results and the Road Ahead.* Policy Research Paper. Washington, DC: World Bank.

———. 1993a. *Implementing the World Bank's Strategy to Reduce Poverty: Progress and Challenges*. Washington, DC: World Bank.

———. 1993b. *Agricultural Sector Review.* Agriculture and Natural Resources Department. Washington, DC: World Bank.

———. 1993c. *A Strategy to Develop Agriculture in Sub-Saharan Africa and a Focus for the World Bank.* World Bank Technical Paper 203. Washington, DC: World Bank.

———. 1992a. *The Effects of Economic Policies on African Agriculture*. World Bank Discussion Papers. Washington, DC: World Bank.

———. 1992b. *An Agricultural Growth and Rural Environment Strategy for the Coastal and Central African Francophone Countries.* Washington, DC: World Bank.

———. 1991a. "Agriculture Technology in Sub-Saharan Africa." A Workshop on Research Issues. World Bank, Washington, DC.

———. 1991b. *Aid to African Agriculture: Lessons from Two Decades of Donors' Experience.* Washington, DC: World Bank.

———. 1991c. *Annual Sector Review: Agriculture and Rural Development.* Washington, DC: World Bank.

———. 1990. *The Long-Term Perspective Study of Sub-Saharan Africa. Background Papers, Volume 2: Economic and Sectoral Policy Issues.* Washington, DC: World Bank.

———. 1989a. *Fertilizer Policy in Africa: Lessons from Development Programs and Adjustment Lending, 1970–87.* MADIA Discussion Paper 8323. Washington, DC: World Bank.

———. 1989b. *Markets, Marketing Boards, and Cooperatives in Africa: Issues in Adjustment Policy*. MADIA Discussion Paper 8436. Washington, DC: World Bank.

———. 1989c. *The Rural Road Question and Nigeria's Agricultural Development.* MADIA Discussion Paper 8437. Washington, DC: World Bank.

———. 1989d. *Cotton in Africa: An Analysis of Differences in Performance.* MADIA Discussion Paper 8297. Washington, DC: World Bank.

———. 1989e. *Growth and Structural Change in East Africa: Domestic Policies, Agricultural Performance and World Bank Assistance, 1963–1986*. MADIA Discussion Paper 8295. Washington, DC: World Bank.

———. 1989f. *Managing Agricultural Development in Africa: Three Articles on Lessons from Experience.* MADIA Discussion Paper 8477. Washington, DC: World Bank.

———. 1989g. *Population Pressure, the Environment and Agricultural Intensification: Variations on the Boserup Hypothesis.* MADIA Discussion Paper 8325. Washington, DC: World Bank.

———. 1989h. *Smallholder and Large-Scale Agriculture in Africa: Are There Tradeoffs between Growth and Equity?* MADIA Discussion Paper 8296. Washington, DC: World Bank.

———. 1989i. *State-Federal Relations in Nigerian Agriculture.* MADIA Discussion Paper 8298. Washington, DC: World Bank.

———. 1989j. *Structural Adjustment, Agricultural Development and the Poor: Lessons from the Malawian Experience,* MADIA Discussion Paper 8324. Washington, DC: World Bank.

———. 1989k. *Agricultural Growth, Domestic Policies, the External Environment, and Assistance to Africa: Lessons of a Quarter Century.* MADIA Discussion Paper 8294. Washington, DC: World Bank.

———. 1987a. "Growth and Structural Change in East Africa: Domestic Policies, Agricultural Performance and World Bank Assistance, 1963–1986." Departmental Working Paper DRD273, World Bank, Washington, DC.

———. 1987b. "Growth and Structural Change in East Africa: Domestic Policies, Agricultural Performance and World Bank Assistance, 1963–1986." Departmental Working Paper DRD274, World Bank, Washington, DC.

———. 1975. *Rural Development.* Sector Policy Paper. Report No. 10272. Washington, DC: World Bank.

Yanggen, David, Valerie Kelly, Thomas Reardon, and Anwar Naseem. 1998. "Incentives for Fertilizer Use in Sub-Saharan Africa: A Review of Empirical Evidence on Fertilizer Response and Profitability." International Development Working Paper No. 70, Michigan State University, East Lansing.